CONCISE COMPLEX ANALYSIS

Revised Edition

CONCISE COMPLEX ANALYSIS

Revised Edition

Sheng Gong
Youhong Gong
University of Science & Technology of China, China

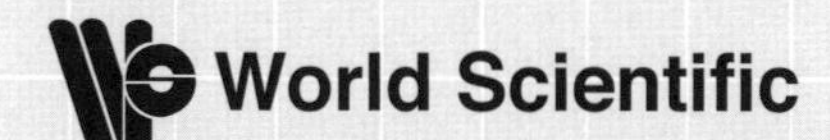

NEW JERSEY · LONDON · SINGAPORE · BEIJING · SHANGHAI · HONG KONG · TAIPEI · CHENNAI

Published by

World Scientific Publishing Co. Pte. Ltd.

5 Toh Tuck Link, Singapore 596224

USA office: 27 Warren Street, Suite 401-402, Hackensack, NJ 07601

UK office: 57 Shelton Street, Covent Garden, London WC2H 9HE

British Library Cataloguing-in-Publication Data
A catalogue record for this book is available from the British Library.

CONCISE COMPLEX ANALYSIS
Revised Edition

ISBN-13 978-981-270-693-5
ISBN-10 981-270-693-3

Printed in Singapore by World Scientific Printers (S) Pte Ltd

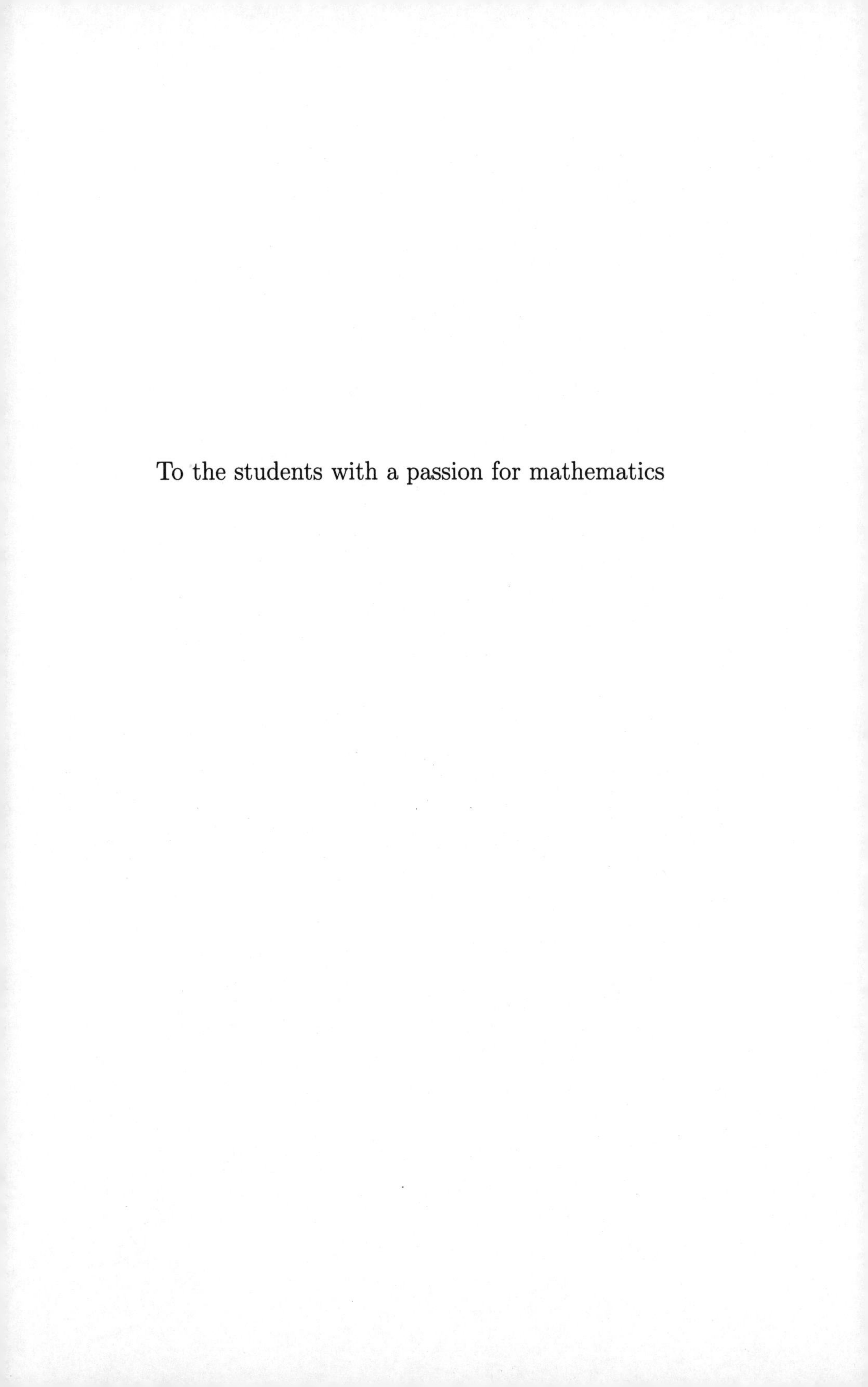

To the students with a passion for mathematics

Preface to the Revised Edition

Two new chapters are added in the revised edition. Chapter 7 covers elliptic functions. Chapter 8 covers the Reimann ζ-function and the Prime Number Theorem. The addition of these chapters is motivated by the many important connections between these functions and numerous other branches of modern mathematics. Many revisions of English language are made in the first six chapters.

We are grateful to Professor Xuean Zheng of Beijing Normal University, China, for the discussions about the materials of this new edition. We thank Professor Der-Chen Chang of Georgetown University, USA, for some useful suggestions and comments after he used this book as the text in his classes.

We also thank Dr. Weiqi Gao for the help he gave in many ways to make the revised edition of this book possible. We especially thank Angela Gao for being the English consultant with her busy schedule.

Finally We would like to express our thanks to Zhang Ji and Jessie Tan of World Scientific for their patience and cooperation during the preparation of this edition.

Sheng Gong
Youhong Gong
September 23, 2006
Beijing

Preface to the First Edition

Professor Gong has written a very appealing book on complex analysis. It is indeed concise but this is far from its only attribute. ("Concise" applies to the discussion of those results in complex variables which are completely analogous to real-variable calculus results, and to Professor Gong's ability to extract the essence of a proof in his presentation.) The book is also insightful, and when an important result can be viewed from different angles, the author rightly feels that it is valuable to point this out. The author's viewpoint is not only that of a geometric function theorist but that of a very broadly trained and broadly published analyst.

There are three noteworthy features of the choice of subject material. First, Cauchy's theorem is treated from the point of view of Green's theorem as well as via the Goursat proof. As well as being pedagogically sound, this permits an introduction to $\bar{\partial}$ techniques wich are used for example in proving the Mittag-Leffler theorem. Second, there is a beautiful chapter entitled "Differential Geometry and Picard's Theorem", which contains a discussion of Gaussian curvature of conformal metrics, the Ahlfors-Schwarz lemma, proofs of Liouville's theorem and Picard's first theorem via the explicit construction of conformal metrics, a geometric discussion of normal families, and Picard's second theorem. Third, there is an introduction to several complex variables which illustrates some of the differences between the one-variable theory and the several-variable theory.

There is more than enough material for a one-semester course, in fact in just 170 pages the reader is taken from the basics to topics which are often part of a second course in complex analysis. There are a great many excellent exercise.

Professor Gong is the student of L.K.Hua and a former Vice-President of the University of Science and Technology of China. He is the author of many books, a number of which have already been translated into English. The English edition of the present text will be a valuable addition to the advanced undergraduate and beginning graduate textbook literature.

Ian Graham
University of Toronto
May, 2000

Foreword

My thought of writing a text book from many years of teaching complex analysis now becomes reality.

There is a long history of studying this subject and many good texts. Since some modern concepts and theorems have been categorized into the basics along the rapid development of mathematics, the idea of writing this book is an attempt to interpret the classical basic materials of complex analysis from the viewpoint of modern mathematics without strictly separating the fundamental knowledge of different branches. In other words, I emphasize the unification of mathematics to the reader by showing the influence and connection between different branches of mathematics. Maybe this attempt will bring some controversies, but I believe this kind of controversy has a good effect on our studies.

Most materials of this book were written the traditional way. The parts which reflect the above purposes are the following:

1. Since complex analysis is calculus on the complex manifold, some of the concepts and theorems in calculus can be generalized to the complex number field without any difficulties. Chapter 1 talks about the most important results of calculus and their generalizations of these results in the complex number field following the ideas of my book "Concise Calculus". The Fundamental Theorem of Calculus on the complex plane is Green's Theorem in complex form. This gives a preparation of the construction of Pompeiu formula in Chapter 2.

2. Traditional complex analysis consists of three parts: Cauchy integral theory, Weierstrass series theory and Riemann geometry theory. These three parts are not generalized from calculus. These traditional materials are covered in Chapters 2, 3 and 4, but I have given modern treatment for some of the theorems.

Cauchy-Green Formula (Pompeiu Theorem) can be derived from Green's Theorem in complex form. The real part and the imaginary part of the function in the formula are C^1 but not necessary holomorphic. Cauchy Integral Theorem and Cauchy Integral Formula are its simple corollaries. The reasons for introducing Pompeiu Formula before Cauchy Integral Theorem and Cauchy Integral Formula are:

(1) It is natural from the viewpoint of complex analysis is calculus on complex number field. The Pompeiu Formula can be derived from Green's Theorem in complex form without any additional conditions but Cauchy Integral Formula cannot.

(2) The solution of one dimensional $\bar{\partial}$-problem can be obtained from Pomeiu Formula but not from Cauchy Integral Formula. It is well known that $\bar{\partial}$-problem is a very important part of partial differential equation theory and it is also a powerful tool of modern mathematics.

In Chapter 3, several theorems are proved by solving the $\bar{\partial}$-problem. The proof of the Mittag-Leffler Theorem by the solution of $\bar{\partial}$-problem is given by me. As the reader can see that the proof becomes very simple when this method is used. I include this proof in this book because I want to show the reader an example of using modern mathematics to treat classical results and how powerful the $\bar{\partial}$-problem is.

In the same chapter, the Interpolation Theorem is also proved by solving the $\bar{\partial}$-problem. Moreover, as another application of the Pompeiu Theorem, the uniform estimation of the derivative of each order of holomorphic functions is given as Theorem 2.8 (2) in Chapter 2. This is a profound theorem and its proof is difficult without using the Pompeiu Formula. The reader can see the importance of the Pompeiu Theorem through this part of discussion.

3. One of the basic theorems of complex analysis introduced in Chapter 3 is Poincare-Koebe Uniformization Theorem: "Any simply connected Riemann surface is one-to-one holomorphically equivalent to one of the following regions: the unit disc, the complex plane $\mathbb{C}$ or the extended complex plane $\mathbb{C}^*$." This is one of the most beautiful theorems in complex analysis. This theorem, along with Abel Theorem and Riemann-Roch Theorem are the three most important theorems in classical Riemann surface theory. This theorem is stated in Section 3.3 of Chapter 3. We omit the proof of the theorem in this book since it is beyond the college level. Besides its many applications, this theorem also points out the importance of the three regions, the unit disc, the complex plane and the extended complex plane in complex analysis, and makes the exploration and discussion of the three

regions an indispensable part of this subject. We can say that it is the most important theorem.

The holomorphic automorphism groups of these three regions are given in Chapters 2 and 3. They are Lie groups in the simplest form. Although the concept of Lie group is not our main concern in this book, letting the reader know some of examples and applications early in order to get ready for their further studies is my intention here. For the same purpose, the holomorphic automorphism groups of the unit ball and the bidisc of $\mathbb{C}^2$ are introduced in Chapter 6, and the classical Poincaré Theorem in several complex variables is proved by using these groups. This gives the reader two more examples of Lie group.

The importance of the holomorphic automorphism group of a region is that it determines some of analysis properties of the region. One of the examples of this kind in this book is the determination of the Poisson kernel by the holomorphic automorphism group of the unit disc.

Just as the three holomorphic automorphism groups in Chapters 2 and 3 represent the relationship between complex analysis and algebra, the contents of Chapter 5 represent the relationship between complex analysis and geometry.

Poincaré-Koebe Theorem can be stated more precisely as:

(1) Any simply connected open hyperbolic Riemann surface can be conformally mapped to a unit disc.

(2) Any simply connected open parabolic Reimann surface can be conformally mapped to the complex plane $\mathbb{C}$.

(3) Any simply connected closed Reimann surface can be conformally mapped to the extended complex plane $\mathbb{C}^*$.

As basic knowledge, the geometry of these three regions is constructed and discussed in Chapter 5 by defining the hyperbolic metric (Poincaré metric), the parabolic metric (Euclidean metric) and the elliptic metric (spherical metric) on the unit disc, $\mathbb{C}$ and $\mathbb{C}^*$ respectively.

Although elementary geometry is used for the proof of some basic facts of complex analysis, differential geometry and complex geometry are what we meant here.

4. There are more such examples in this book, like the Schwarz-Pick Lemma in Chapter 2: "The Poincaré metric does not increase under a holomorphic mapping from the unit disc to the unit disc." This lemma gives the classic Schwarz Lemma a natural geometric explanation. The famous Picard Theorem which is proved by geometric methods in Chapter 5 is another such example.

In 1938, Lars V. Ahlfors proved the Alhfors-Schwarz Lemma which marks the entrance of modern differential geometry into complex analysis. The proof of the Picard Theorem is not included in some of the traditional text books due to the requirement of advanced knowledge—elliptic modular functions. In Chapter 5 of this book, Picard Theorem is proved through the Alhfors-Schwarz Lemma. The reader can learn this theorem completely with its proof as well as to see the power of modern mathematics and the connection between different branches of mathematics.

Another example of this kind of connection is the Poisson Integral Formula in Section 2.6 Chapter 2. The traditional analytical approach is not used in this book because I want the reader to see how wonderful the Lie group approach is by the clean proof of Poisson Formula through the holomorphic automorphism group of the unit disc.

From the viewpoint of complex analysis, the Poisson formula is an integral representation; from the viewpoint of partial differential equation, it is the solution of Dirichlet problem of Laplace equations, and from the viewpoint of harmonic analysis, it is the Abel sum of the Fourier series of a function. Although this is a simple example of the connection between different branches of mathematics, it is very important. It also shows that a result from one subject can leads to a result in another subject. The other example of this kind is the Riemann surface. From the viewpoint of complex analysis, the Riemann surface is a one dimensional complex manifold, more precisely, a one dimensional Kähler manifold; from the viewpoint of algebraic geometry, it is an algebraic curve; and from the viewpoint of algebraic number theory, it is a field of algebraic functions with one variable. Any mathematical theory that connects different branches carries the potential of new development of mathematics.

5. In the last chapter, Chapter 6, we talk about complex analysis in several variables. The purpose of this chapter is not to teach this subject, instead, it is to let the reader understand complex analysis better through some of the main theorems. Poincaré Theorem and Hartogs Theorem were discussed in this chapter. Poincaré Theorem states: "The unit ball is not holomorphically equivalent to the bidisc in $\mathbb{C}^n$." (We only talk about the case of $\mathbb{C}^2$ in this book for simplicity.) This theorem tells us that Riemann Mapping Theorem (topologically equivalence implies holomorphic equivalence) only holds in complex analysis of one variable. It does not hold in calculus and does not hold in complex analysis in several variables either. It is a profound theorem in complex analysis. The theory of geometry in complex analysis also exists only in the case of one variable. It can not

be generalized to higher dimensions directly. The other theorem, Hartogs Theorem states: "Suppose $\Omega \subseteq \mathbb{C}^n$ $(n \geq 2)$ is a region, K is a compact subset of Ω and $\Omega \setminus K$ is connected. If f is holomorphic on $\Omega \setminus K$, then f has an analytic continuation on Ω." This theorem tells us that, in general, the theory of Laurent series which is the main part of Weierstrass series theory does not exists in complex analysis of several variables. It follows that the related topics such as the definition of meromorphic functions, the discussion of singularity by series can not be generalized to the case of several complex variables directly. As we can see, this is also a profound theorem. Therefore, as the starting points of two of the three main parts of complex analysis, Riemann Mapping Theorem and Laurent series can not be generalized to several complex variables. I think the purpose of this chapter can be fulfilled by the discussions of these two theorems.

6. Besides to bring some convenience to the reader, the three appendices emphasize the most basic and important materials of modern mathematics and the connection of different branches of mathematics is mentioned again.

From all six parts above, it is easy to see why I prefer to use "Complex Analysis" as the name of this book instead of "The Function Theory of One Complex Variable" or "The Theory of Holomorphic Functions". The function is not the only topic we talk about!

I believe the reader will find this book beneficial to their studies.

Sheng Gong
May 31, 2000
Beijing

Contents

Chapter 1

Calculus

1.1 A Brief Review of Calculus

Complex analysis is calculus on the field of complex numbers. Similar to the generalizations of any other branches of mathematics, parts of the material of calculus can be generalized to the complex number field without difficulties while other parts are different from the original theory on the field of real numbers. It is this difference that attracts the interest of mathematicians, since it reveals the characteristics of complex analysis.

In this chapter, we give a brief review of calculus and discuss results that can be directly generalized to the field of complex numbers. We will discuss properties and results that are unique to the field of complex numbers in the following chapters.

The three main parts of calculus are differentiation, integration and the fundamental theorem of calculus which connections the first two. Known as the Newton-Leibniz formula, this theorem demonstrates the inverse relationship between differentiation and integration.

It is well known that if $y = f(x)$ is a function on a interval (a, b), and if

$$\lim_{h \to 0} \frac{f(x+h) - f(x)}{h}$$

exists for a point x in (a, b), then we say that $f(x)$ is *differentiable* at x. The limit, denoted df/dx or $f'(x)$ is called the *derivative* of $f(x)$ at x. Also, $df = f'(x)dx$ is called the *differential* of $f(x)$ at x. If $f(x)$ is differentiable at every point of (a, b), then $f(x)$ is differentiable on (a, b). On the other hand, let $y = f(x)$ be defined on $[a, b]$, and let $x_i \in [a, b]$, $1 \le i \le n$, be such that $a = x_0 < x_1 < \cdots < x_n = b$ and the length of each $[x_{i-1}, x_i]$,

$(i = 1, \cdots, n)$, tends to zero when n tends to ∞. If

$$\lim_{n\to\infty} \sum_{i=1}^{n} f(\xi_i)(x_i - x_{i-1})$$

exists for $\xi_i \in [x_{i-1}, x_i]$, then we say that $f(x)$ is *integrable* on $[a, b]$. This limit, denoted

$$\int_a^b f(x)\,dx$$

is called the *integral* of $f(x)$ on $[a, b]$. These are the basic definitions and the starting points of calculus. They have clear geometric meanings: the derivative is the slope of the tangent line of $f(x)$ at the point $(x, f(x))$ and the integral is the area formed by $f(x)$, x-axis, $y = a$ and $y = b$.

The fundamental theorem of calculus plays the most important role in calculus. There are two equivalent forms of this theorem:

Theorem 1.1 ***(The Fundamental Theorem of Calculus) (differential form)*** *Suppose the function $f(x)$ is continuous on $[a, b]$ and $x \in [a, b]$. Let*

$$\Phi(x) = \int_a^x f(t)dt \qquad (a \le x \le b)$$

Then $\Phi(x)$ is differentiable in $[a, b]$ and $\Phi'(x) = f(x)$, $d\Phi(x) = f(x)dx$. In other words, if the integral of $f(x)$ is $\Phi(x)$, then the differential of $\Phi(x)$ is $f(x)dx$.

Theorem 1.2 ***(The Fundamental Theorem of Calculus) (integral form)*** *Suppose $\Phi(x)$ is differentiable in $[a, b]$, and $d\Phi(x)/dx$ is equal to a continuous function $f(x)$. Then*

$$\int_a^x f(t)dt = \Phi(x) - \Phi(a) \qquad (a \le x \le b)$$

holds. In other words: If the differential of $\Phi(x)$ is $f(x)dx$, then the integral of $f(x)$ is $\Phi(x)$.

By this theorem, finding the integral is the inverse operation of finding the differential of a function, and properties of differentials correspond to properties of integrals. For instance:

$$\frac{d(f(x) \pm g(x))}{dx} = \frac{df(x)}{dx} \pm \frac{dg(x)}{dx}$$

corresponds to

$$\int (f(x) \pm g(x))dx = \int f(x)dx \pm \int g(x)dx;$$

$$\frac{d}{dx}(fg) = f\frac{dg}{dx} + \frac{df}{dx}g$$

corresponds to

$$\int fg'dx = fg - \int gf'dx.$$

If $u = f(y)$, $y = g(x)$, then

$$\frac{df(g(x))}{dx} = \frac{df}{dy} \cdot \frac{dy}{dx}$$

corresponds to

$$\int f(g(x))g'(x)dx = \int f(y)dy$$

etc.

Another correspondence is between the mean value theorems:

Theorem 1.3 ***(Mean Value Theorem for Derivatives)*** *If $f(x)$ is differentiable on $[a, b]$, then there exists $c \in [a, b]$ such that*

$$f(b) - f(a) = f'(c)(b - a).$$

Theorem 1.4 ***(Mean Value Theorem for Integrals)*** *If $f(x)$ is continuous on $[a, b]$, then there exists $\xi \in [a, b]$ such that*

$$\int_a^b f(x)dx = f(\xi)(b - a).$$

Moreover, the Taylor expansion of a function can be proved by differentiation and by integration. The remainder of a Taylor series can be expressed by the differential form and by the integral form as well.

Elementary functions and their compositions are discussed in calculus. The following three classes of functions are what we mean by elementary functions:

1. Power functions x^α, where α is a real number; polynomials $a_0 + a_1x + \cdots + a_nx^n$, where a_i $(i = 0, 1, \cdots, n)$ are constants; rational fractions

$$\frac{b_0 + b_1x + \cdots + b_mx^m}{c_0 + c_1x + \cdots + c_px^p},$$

where b_i $(i = 0, 1, \cdots, m)$, c_i $(i = 0, 1, \cdots, p)$ are constants; and their inverse functions.

2. Trigonometric functions $\sin x$, $\cos x$, $\cdots$, and their inverse functions $\arcsin x$, $\arccos x$, $\cdots$.

3. Exponential functions e^x, 2^x, $\cdots$, and their inverse functions $\ln x$, $\log_2 x$, $\cdots$.

The Taylor series for a function $f(x)$ is an approximation of $f(x)$ by functions in the first class. The Fourier series for a function $f(x)$ is an approximation of $f(x)$ by $\sin nx$ and $\cos nx$, $(n = 0, 1, 2, \cdots)$, in the second class. One of the reasons that there is no approximation of $f(x)$ by exponential functions is that exponential functions can be expressed by trigonometric functions according to the Euler formula. The Taylor series for the following important elementary functions are well known:

$$e^x = 1 + \frac{x}{1!} + \frac{x^2}{2!} + \frac{x^3}{3!} + \cdots, \tag{1.1}$$

$$\sin x = \frac{x}{1!} - \frac{x^3}{3!} + \frac{x^5}{5!} - \frac{x^7}{7!} + \cdots, \tag{1.2}$$

$$\cos x = 1 - \frac{x^2}{2!} + \frac{x^4}{4!} - \frac{x^6}{6!} + \cdots, \tag{1.3}$$

$$\ln(1+x) = x - \frac{x^2}{2} + \frac{x^3}{3} - \cdots \qquad (-1 < x \le 1), \tag{1.4}$$

$$(1+x)^r = 1 + rx + \frac{r(r-1)}{2!}x^2 + \frac{r(r-1)(r-2)}{3!}x^3 + \cdots$$

$$(|x| < 1, \text{r is a real number}). \tag{1.5}$$

The above discussion is a brief outline of calculus. The reader is referred to the related parts of *Concise Calculus* by Sheng Gong and Shenglei Zhang (Sheng Gong, Shenglei Zhang [1]) for more details.

There are also three parts in calculus of multiple variables: differentiation, integration and the fundamental theorem of calculus which connects the first two. The differentiation part includes the partial differential, the total differential and the Jacobi matrix which corresponds to the derivative in one variable. The integration part includes the multiple integral, the line integral and the surface integral. Since these are the natural generalizations of derivatives and integrals in one variable, there are also corresponding theorems. We shall not list them here. In the case of higher dimensions, the Green's theorem, the Stokes's theorem and the Gauss's theorem demonstrate the inverse relation between the differential and the integral. They take the place of the fundamental theorem of calculus.

Theorem 1.5 ***(Green's Theorem)*** *Let $P(x,y)$ and $Q(x,y)$ be functions that have continuous first partial derivatives on a simply connected region D with a piecewise smooth boundary L. Then*

$$\oint_L Pdx + Qdy = \iint\limits_D \left(\frac{\partial Q}{\partial x} - \frac{\partial P}{\partial y}\right) dx\, dy. \tag{1.6}$$

Theorem 1.6 ***(Stokes's Theorem)*** *Let $P(x,y,z)$, $Q(x,y,z)$ and $R(x,y,z)$ be functions that have continuous first partial derivatives on a surface Σ bounded by a piecewise smooth simple closed curve L. Then*

$$\begin{aligned}&\oint_L Pdx + Qdy + Rdz \\ &= \iint\limits_\Sigma \left(\frac{\partial R}{\partial y} - \frac{\partial Q}{\partial z}\right) dy\, dz + \left(\frac{\partial P}{\partial z} - \frac{\partial R}{\partial x}\right) dz\, dx + \left(\frac{\partial Q}{\partial x} - \frac{\partial P}{\partial y}\right) dx\, dy.\end{aligned} \tag{1.7}$$

Theorem 1.7 ***(Gauss's Theorem)*** *Let $P(x,y,z)$, $Q(x,y,z)$ and $R(x,y,z)$ be functions that have continuous first partial derivatives on a solid region V bounded by a closed surface Σ. Then*

$$\oiint\limits_\Sigma P\, dy\, dz + Q\, dz\, dx + R\, dx\, dy = \iiint\limits_V \left(\frac{\partial P}{\partial x} + \frac{\partial Q}{\partial y} + \frac{\partial R}{\partial z}\right) dV. \tag{1.8}$$

These three theorems describe the relationship between the integral on the boundary of a region and the integral on the interior of a region. They can be written as one theorem referred to as the Stokes's Theorem by using exterior differential forms. We omit the detailed treatment of exterior differential forms here since it can be readily found in many calculus text books. My book *Concise Calculus* was written with this point of view. (Sheng Gong, Shenglei Zhang [1])

We now briefly introduce the exterior differential form of three dimensional Euclidean space. Define the *exterior product* of differentials dx and dy as $dx \wedge dy$. It has the following properties:

(1) $dx \wedge dx = 0$, the exterior product of two same differentials is zero.

(2) $dx \wedge dy = -dy \wedge dx$, the exterior product of two different differentials will have the opposite sigh when the order is transposed.

Of course, (1) can be considered as a corollary of (2). The exterior product of differentials multiplied by a function becomes a differential form, and it is called an *exterior differential form*. For instance: if P, Q, R, A, B, C, H are functions of x, y, z, then

$$Pdx + Qdy + Rdz$$

is an exterior differential form of degree one; (This is the same as the ordinary differential form since it doesn't contain exterior products.)

$$Adx \wedge dy + Bdy \wedge dz + Cdz \wedge dx$$

is an exterior differential form of degree two;

$$Hdx \wedge dy \wedge dz$$

is an exterior differential form of degree three. And P, Q, R, A, B, C, H are called the *coefficients of the differential forms.*

We can define the *exterior differential operator* d for exterior differential form ω as follows:

For an exterior differential form of degree zero, which is a function f, define

$$df = \frac{\partial f}{\partial x}dx + \frac{\partial f}{\partial y}dy + \frac{\partial f}{\partial z}dz.$$

This is just the ordinary total differential operator. For the exterior differential form of degree one $\omega = Pdx + Qdy + Rdz$, define

$$d\omega = dP \wedge dx + dQ \wedge dy + dR \wedge dz.$$

That is, take differentials of P, Q, R, then perform the exterior product. By the properties of exterior product, we have

$$d\omega = (\frac{\partial R}{\partial y} - \frac{\partial Q}{\partial z})dy \wedge dz + (\frac{\partial P}{\partial z} - \frac{\partial R}{\partial x})dz \wedge dx + (\frac{\partial Q}{\partial x} - \frac{\partial P}{\partial y})dx \wedge dy$$

Similarly, for the exterior differential form of degree two, $\omega = Ady \wedge dz + Bdz \wedge dx + Cdx \wedge dy$, define

$$\begin{aligned} d\omega &= dA \wedge dy \wedge dz + dB \wedge dz \wedge dx + dC \wedge dx \wedge dy \\ &= (\frac{\partial A}{\partial x} + \frac{\partial B}{\partial y} + \frac{\partial C}{\partial z}) dx \wedge dy \wedge dz; \end{aligned}$$

for the exterior differential form of degree three, $\omega = Hdx \wedge dy \wedge dz$, define

$$d\omega = dH \wedge dx \wedge dy \wedge dz,$$

and this is equal to zero.

If we let $ddx = ddy = ddz = 0$, then the exterior differential operator d is the same as the ordinary differential operator. That is, the exterior differential operator is applied to each term, and, within each term, applied to each factor in turn while keeping the rest of the factor unchanged. The only difference is that we use exterior product instead of regular product when exterior differential operator is applied. We have the important Poincaré lemma:

Theorem 1.8 ***(Poincaré Lemma)*** *If ω is an exterior differential form, and if the coefficients of ω have continuous partial derivatives of second order, then $dd\omega = 0$. The converse is also true, if ω is an exterior differential form of degree p, and $d\omega = 0$, then there exists an exterior differential form α of degree $p - 1$, such that $\omega = d\alpha$.*

Now we are ready to write the Green's theorem, the Stokes's theorem and the Gauss's theorem in one equation:

$$\int_{\partial\Sigma} \omega = \int_{\Sigma} d\omega, \tag{1.9}$$

where ω is an exterior differential form, $d\omega$ is the exterior differential of ω, Σ is the closed integral region of $d\omega$, $\partial\Sigma$ is the boundary of Σ, the multiplicity of the integral is the same as the dimension of the region. In fact, (1.9) is just Newton-Leibniz Theorem when the degree of the exterior differential form ω is zero. It is the Green's Theorem in two dimensional Euclidean space when the degree of ω is one, and is the Stokes's Theorem in three dimensional Euclidean space. It is the Gauss's Theorem when the degree of ω is two. Equation (1.9) perfectly demonstrates the inverse relation between differentiation and integration. It holds not only in the three dimensional Euclidean space but also in the Euclidean space with arbitrarily high dimensions. Moreover, it also holds on differential manifolds.

Therefore, equation (1.9) is the fundamental theorem of calculus in high dimensional spaces, and this theorem is the apex of calculus.

This is of course a very brief review of calculus. I think it is enough to understand the train of thought clearly and it is not necessary to state more details here.

Since complex analysis is calculus on complex number field and the continuation of calculus, it is natural that equation (1.9) becomes one of the starting points of this book.

1.2 The Field of Complex Numbers, The Extended Complex Plane and Its Spherical Representation

All complex numbers form a field that is an extension of the real number field.

It is well known in elementary algebra that the imaginary unit i has the property $i^2 = -1$. Combining with real numbers α and β by addition and multiplication, we obtain a complex number $\alpha + i\beta$, where α and β are the real and imaginary parts. Let $a = \alpha + i\beta$. Then we denote $\operatorname{Re} a = \alpha$ and $\operatorname{Im} a = \beta$. Two complex numbers are equal if and only if their real and imaginary parts are equal respectively. The ordinary laws of arithmetic concerning addition, subtraction, multiplication and division are:

If $\alpha + i\beta$ and $\gamma + i\delta$ are complex numbers, then

$$(\alpha + i\beta) \pm (\gamma + i\delta) = (\alpha \pm \gamma) + (\beta \pm \delta),$$

$$(\alpha + i\beta)(\gamma + i\delta) = (\alpha\gamma - \beta\delta) + i(\alpha\delta + \beta\gamma),$$

and if $\gamma + i\delta \neq 0$, then

$$\frac{\alpha + i\beta}{\gamma + i\delta} = \frac{(\alpha + i\beta)(\gamma - i\delta)}{(\gamma + i\delta)(\gamma - i\delta)} = \frac{(\alpha\gamma + \beta\delta) + i(\beta\gamma - \alpha\delta)}{\gamma^2 + \delta^2}.$$

If $a = \alpha + i\beta$, then $\alpha - i\beta$ is called the *conjugate* of a, and denoted by $\bar{a}$. Thus, we have

$$\operatorname{Re} a = \frac{a + \bar{a}}{2}, \quad \operatorname{Im} a = \frac{a - \bar{a}}{2i},$$

$$\overline{a + b} = \bar{a} + \bar{b}, \quad \overline{ab} = \bar{a} \cdot \bar{b}, \quad \overline{\left(\frac{a}{b}\right)} = \frac{\bar{a}}{\bar{b}}, \quad a\bar{a} = \alpha^2 + \beta^2.$$

We denote $|a|^2 = a\bar{a}$. And $|a| = \sqrt{\alpha^2 + \beta^2}$ is called *the absolute value of* a. Obviously,

$$|a| \geq 0, \quad |ab| = |a| \cdot |b|, \quad \left|\frac{a}{b}\right| = \frac{|a|}{|b|} \quad (b \neq 0),$$

$$|a \pm b|^2 = |a|^2 + |b|^2 \pm 2\,\mathrm{Re}\, a\bar{b}, \quad |a + b| \leq |a| + |b|.$$

A complex number $a = \alpha + i\beta$ can be represented as a point with coordinates (α, β) in a rectangular coordinate system on a plane. The first coordinate axis is called the *real axis*, and the second coordinate axis is called the *imaginary axis*. The plane itself is called the *complex plane* and is denoted by $\mathbb{C}$.

A complex number not only can be represented as a point but also can be represented as a vector from the origin to the point. The same symbol a denotes the complex number, the point and the vector. Since all vectors which can be obtained by a parallel displacement of a vector are considered the same, the addition of complex numbers becomes the addition of vectors. The formulas about complex numbers are often given geometric meanings. For example, $|a|$ is the length of the vector a; $|a+b| \leq |a| + |b|$ means that the sum of two sides of a triangle is greater than or equal to the third side.

Polar coordinates (r, φ) can also be used to represent complex numbers. Let $a = \alpha + i\beta = r(\cos\varphi + i\sin\varphi)$. Then $r = |a|$ is called the *modulus* of a and φ is called the *argument* of a. If

$$a_1 = r_1(\cos\varphi_1 + i\sin\varphi_1), \;\; a_2 = r_2(\cos\varphi_2 + i\sin\varphi_2),$$

then

$$\begin{aligned} a_1 a_2 &= r(\cos\varphi + i\sin\varphi) \\ &= r_1 r_2(\cos\varphi_1 + i\sin\varphi_1)(\cos\varphi_2 + i\sin\varphi_2) \\ &= r_1 r_2(\cos(\varphi_1 + \varphi_2) + i\sin(\varphi_1 + \varphi_2)). \end{aligned}$$

This implies that $r = r_1 r_2$ and $\varphi = \varphi_1 + \varphi_2$. The argument of a complex number a is not unique since $\varphi + 2k\pi$, where k is an integer, are also arguments of a. We denote the arguments of a by $\mathrm{Arg}\, a$. Especially, φ is called the *principle argument* of a denoted by $\arg a$, if $0 \leq \varphi < 2\pi$.

Let $z = x + iy$, a be a complex number and r be a real number. Then $|z - a| = r$ is a circle with center a and radius r; $|z - a| < r$ is a disc with center a and radius r. We denote this disc by $D(a, r)$. Similarly, the upper half plane is represented by $\operatorname{Im} z > 0$, and the right half plane is represented by $\operatorname{Re} z > 0$.

The infinity point, denoted ∞, extends the complex plane. For every finite complex number $a \in \mathbb{C}$, we have $a + \infty = \infty + a = \infty$ and $a/0 = \infty\,(a \neq 0)$. For every $b \neq 0$ we have $b \cdot \infty = \infty \cdot b = \infty$ and $b/\infty = 0$. All the points in $\mathbb{C}$ together with "∞" form the extended complex plane, denoted by $\mathbb{C}^*$. Thus $\mathbb{C}^* = \mathbb{C} \bigcup \infty$.

We now introduce the stereographic projection to construct a geometric model on which every point of the extended complex plane has a representation.

Consider the unit sphere $S^2 = \{(x_1, x_2, x_3) \in \mathbb{R}^3 | x_1^2 + x_2^2 + x_3^2 = 1\}$. For every point on S^2 except $(0, 0, 1)$, there is a corresponding complex number

$$z = \frac{x_1 + ix_2}{1 - x_3}. \tag{1.10}$$

This correspondence is one-to-one. In fact, it follows from (1.10) that

$$|z|^2 = \frac{x_1^2 + x_2^2}{(1 - x_3)^2} = \frac{1 - x_3^2}{(1 - x_3)^2} = \frac{1 + x_3}{1 - x_3}.$$

Thus, we have

$$x_1 = \frac{z + \bar{z}}{1 + |z|^2}, \quad x_2 = \frac{z - \bar{z}}{1 + |z|^2}, \quad x_3 = \frac{|z|^2 - 1}{|z|^2 + 1}. \tag{1.11}$$

Let the infinity point ∞ correspond to $(0, 0, 1)$. Then the one-to-one correspondence between the points on the sphere S^2 and the extended complex plane $\mathbb{C}^*$ is completed. Therefore, the sphere S^2 can be considered as a representation of the extended complex plane $\mathbb{C}^*$. The sphere S^2 is called the *Riemann sphere*. Obviously, the hemisphere $x_3 < 0$ corresponds to the inside of the unit disc, $|z| < 1$, and the hemisphere $x_3 > 0$ corresponds to the outside of the unit disc, $|z| > 1$.

The geometric meaning of (1.10) is clear in the complex plane with x_1-axis and x_2-axis as its real and imaginary axes respectively.

Let $z = x + iy$. Then, by (1.10), we have

$$x : y : -1 = x_1 : x_2 : x_3 - 1.$$

This shows that the points $(x, y, 0)$, (x_1, x_2, x_3), $(0, 0, 1)$ are collinear. Thus, the correspondence is actually a central projection with center $(0, 0, 1)$, and projects the points of S^2 onto $\mathbb{C}^*$. This projection is called *stereographic* *projection*. The infinity point is no longer special in the spherical representation.

1.3 Derivatives of Complex Functions

As in calculus, a complex valued function $w = f(z)$ is defined on the complex field, where z and w are complex numbers. We restrict $f(z)$ to be single valued for the moment in order to have a clear definition. We define the limit of function $f(z)$ as

$$\lim_{z\to a} f(z) = A$$

if for any given $\varepsilon > 0$, there exists a $\delta > 0$ such that $|f(z)-A| < \varepsilon$ whenever $|z-a| < \delta$ and $z \neq a$. A function is continuous at $z = a$ if

$$\lim_{z\to a} f(z) = f(a).$$

As with calculus, we can define open sets, closed sets, connected sets and compact sets on the complex plane. A curve on the complex plane is defined by a complex valued continuous function $\gamma(t)$ on an interval $[\alpha, \beta]$, where $\gamma(t) = x(t) + iy(t)$, $\alpha \leq t \leq \beta$. The functions $x(t)$ and $y(t)$ are real valued continuous functions of t. The end points of the curve $\gamma(t)$ are $\gamma(\alpha)$ and $\gamma(\beta)$. The curve is closed if $\gamma(\alpha) = \gamma(\beta)$. The direction of the curve is the direction along which t is increasing. If $\gamma'(t)$ exists and is continuous, then $\gamma(t)$ is called a *smooth curve*. If $\gamma'(t)$ is continuous except on a finite number of points, and $\gamma(t)$ has left and right derivatives at those points, then $\gamma(t)$ is called *piecewise smooth*. A piecewise smooth curve is rectifiable. A curve $\gamma(t)$ is called a *simple curve* or a *Jordan curve* if $\gamma(t_1) = \gamma(t_2)$ only if $t_1 = t_2$. If $\gamma(t)$ is also a closed curve, then it is called a *simple closed curve* or a *Jordan closed curve*.

A set D on the complex plane is called a *region*, if

(1) D is an open set;

(2) D is connected, that is, any two points in D can be joined by a curve completely contained in D.

We shall omit the complicated proof of the following "obvious" fact.

Theorem 1.9 *(Jordan Theorem) The complex plane can be divided into two regions by a simple closed curve. One of the regions is bounded and is called the inside of γ. The other region is unbounded and is called the outside of γ. The curve γ is the common boundary of the two regions.*

The boundary of a region D is denoted by ∂D. If the inside of any simple closed curves in D is completely contained in D, then D is *simply connected*. A connected region is *multiply connected* if it is not simply connected. A region bounded by two non-intersecting Jordan closed curves is two-connected, and a region bounded by n non-intersecting Jordan closed curves is n-connected. These closed curves can degenerating to a single point or a slit. Moreover, the Heine-Borel theorem and the Bolzano-Weierstrass theorem can also be proved just as in calculus. We shall omit the proofs of these theorems here.

Theorem 1.10 ***(Heine-Borel Theorem)*** *If A is a compact set and G is an open covering of A, then G has a finite subcovering.*

Theorem 1.11 ***(Bolzano-Weierstrass Theorem)*** *An infinite subset of a compact set must have a limit point.*

We now investigate the derivatives of complex valued functions of a complex variable. Let $w = f(z)$. Then, naturally, we consider

$$\lim_{h\to 0} \frac{f(z+h) - f(z)}{h},$$

where h is a complex number. If for all different paths that h could approach zero, the limit exists and remains the same, then we say that $f(z)$ is *differentiable* at z, and the limit is denoted by df/dz or $f'(z)$. This limit is called the *derivative* of $f(z)$ at z. If $f(z)$ is differentiable at every point of its domain, then $f(z)$ is called an *analytic function* or a *holomorphic function* on its domain. This definition is the same as the definition of derivative in calculus. Hence, the arithmetic rules for derivatives and the chain rule are still valid. The reader should be able to write these formulas out without any difficulties. There are some differences between the derivatives of complex functions and the derivatives of real functions. If

$$f(z) = u(z) + iv(z) = u(x,y) + iv(x,y)$$

is differentiable at the point $z_0 = x_0 + iy_0$, then

$$\lim_{z\to z_0} \frac{f(z) - f(z_0)}{z - z_0} = f'(z_0)$$

exists and is the same, regardless of the way z approaches z_0. It follows that this limit should exist and is the same for z approaching z_0 through the paths parallel to the coordinate axes. First, let $z = x + iy_0$ and $x \to 0$.

Then

$$f'(z_0) = \lim_{x \to x_0} \left[\frac{u(x, y_0) - u(x_0, y_0)}{x - x_0} + i \frac{v(x, y_0) - v(x_0, y_0)}{x - x_0} \right]$$
$$= u_x(x_0, y_0) + i v_x(x_0, y_0).$$

Next, let $z = x_0 + iy$ and $y \to 0$. Then

$$f'(z_0) = \lim_{y \to y_0} \left[\frac{u(x_0, y) - u(x_0, y_0)}{i(y - y_0)} + \frac{v(x_0, y) - v(x_0, y_0)}{y - y_0)} y - y_0 \right]$$
$$= v_x(x_0, y_0) - i u_y(x_0, y_0),$$

where u_x, u_y, v_x, v_y represent the partial derivative of u and v with respect to x, y. Comparing the real and the imaginary parts of the two equations, we get that

$$u_x = v_y, \quad u_y = -v_x \tag{1.12}$$

at the point (x_0, y_0). These equations can also be written as

$$\frac{\partial f}{\partial x} = -i \frac{\partial f}{\partial y} \tag{1.13}$$

Both (1.12) and (1.13) are called *Cauchy-Riemann equations* or *C-R equations.* Satisfying these equations is a necessary condition for $f(x)$ to be differentiable at a point $z = z_0$, but not a sufficient condition. For example

$$f(z) = f(x + iy) = \sqrt{|xy|}$$

satisfies C-R equation but is not differentiable at the point $z = 0$ by the following argument. Let $x = \alpha t$ and $y = \beta t$. Then

$$\frac{f(z) - f(0)}{z - 0} = \frac{f(z)}{z} = \frac{\sqrt{|\alpha\beta|}}{\alpha + i\beta}.$$

The limit is not unique when $z \to 0$. We have the following theorem.

Theorem 1.12 *The necessary and sufficient condition for a function $f(z) = u + iv$ to be holomorphic on a region D is that u and v have first order continuous partial derivatives on D and satisfy C-R equations (1.12).*

Proof (necessity) If $f(z)$ is differentiable at $z = z_0$, then (1.12) holds. (We will show that the derivative of an analytic function is also an analytic function in Section 2.3 of next chapter. Hence $f' = u_x + iv_x = v_y - iu_y$ is continuous as well.)

(sufficiency) Assume that u and v have first order continuous partial derivatives at point $z_0 = x_0 + iy_0$ and satisfy C-R equations. Let $\alpha = u_x(x_0, y_0)$ and $\beta = v_x(x_0, y_0)$. Then

$$u(x,y) - u(x_0,y_0) = \alpha(x - x_0) - \beta(y - y_0) + \varepsilon_1(|\Delta z|),$$

$$v(x,y) - v(x_0,y_0) = \beta(x - x_0) + \alpha(y - y_0) + \varepsilon_2(|\Delta z|),$$

where $|\Delta z| = \sqrt{(x-x_0)^2 + (y-y_0)^2}$ and ε_1, ε_2 satisfy

$$\lim_{|\Delta z| \to 0} \frac{\varepsilon_1(|\Delta z|)}{|\Delta z|} = \lim_{|\Delta z| \to 0} \frac{\varepsilon_2(|\Delta z|)}{|\Delta z|} = 0$$

Multiplying i to the second equation and adding to the first, we get

$$f(z) - f(z_0) = (\alpha + i\beta)(z - z_0) + \varepsilon_1(|\Delta z|) + i\varepsilon_2(\Delta z|).$$

Dividing by $z - z_0$ on both sides of above equation, we get

$$\frac{f(z) - f(z_0)}{z - z_0} - (\alpha + i\beta) = \frac{\varepsilon_1(|\Delta z|) + i\varepsilon_2(|\Delta z|)}{z - z_0}.$$

Thus,

$$\lim_{z \to z_0} \frac{f(z) - f(z_0)}{z - z_0} = \alpha + i\beta.$$

Therefore,

$$f'(z_0) = u_x(x_0, y_0) + iv_x(x_0, y_0),$$

and the theorem is proved.

If $f(z) = u + iv$ is holomorphic in a region D, then the derivative of $f(z)$ is also a holomorphic function in D. Hence the second order partial derivatives of u and v are also continuous. Therefore, the second order mixed partial derivatives $\partial^2 u/\partial x \partial y$ and $\partial^2 v/\partial y \partial x$ are equal. We will prove this fact in Section 2.3 of next chapter, but use it now in the following statement.

By the C-R equations, we have

$$\frac{\partial^2 u}{\partial x^2} = \frac{\partial^2 v}{\partial x \partial y}, \quad \frac{\partial^2 u}{\partial y^2} = -\frac{\partial^2 v}{\partial y \partial x}.$$

Thus

$$\frac{\partial^2 u}{\partial x^2} + \frac{\partial^2 u}{\partial y^2} = 0.$$

Similarly, we have

$$\frac{\partial^2 v}{\partial x^2} + \frac{\partial^2 v}{\partial y^2} = 0.$$

This type of equation is called the *Laplace equation*. It is one of the basic equations in the theory of partial differential equations—a typical equation of elliptic type, and is denoted by

$$\Delta u = \frac{\partial^2 u}{\partial x^2} + \frac{\partial^2 u}{\partial y^2} = 0,$$

where

$$\Delta = \frac{\partial^2}{\partial x^2} + \frac{\partial^2}{\partial y^2}.$$

A function u is called a *harmonic function* if it satisfies $\Delta u = 0$. The real part and the imaginary part of a holomorphic function $f = u + iv$ are harmonic functions.

Let $z = x + iy$ and $\bar{z} = x - iy$. Then we have

$$x = \frac{1}{2}(z + \bar{z}), \ \ y = -\frac{1}{2}i(z - \bar{z}).$$

So a function of x and y, $f(x, y)$, can be considered as a function of z and $\bar{z}$, $f(z, \bar{z})$ (we ignore the fact that they are conjugate to each other and consider them as two independent variables here).

By the rules of derivative, we have

$$\frac{\partial f}{\partial z} = \frac{1}{2}\left(\frac{\partial f}{\partial x} - i\frac{\partial f}{\partial y}\right), \quad \frac{\partial f}{\partial \bar{z}} = \frac{1}{2}\left(\frac{\partial f}{\partial x} + i\frac{\partial f}{\partial y}\right).$$

This implies that, a function is holomorphic if and only if $\partial f/\partial \bar{z} = 0$. In other words, a holomorphic function is independent of $\bar{z}$. It is only a function of z. Thus, holomorphic functions can be considered as functions of a single complex variable z instead of complex valued functions of two real variables.

The equation $\partial f/\partial \bar{z} = 0$ is equivalent to the equation $\partial f/\partial x = \partial f/\partial z$. So, we have

$$\Delta = 4\frac{\partial}{\partial z}\frac{\partial}{\partial \bar{z}} = 4\frac{\partial}{\partial \bar{z}}\frac{\partial}{\partial z}.$$

An important geometric property of complex derivatives is conformality.

Consider a holomorphic function $f(z)$ in a region D, a point z_0 in D with $f'(z_0) \neq 0$. Suppose that $\gamma(t)\,(0 \leq t \leq 1)$ is a smooth curve in D

passing through z_0 with $\gamma(0) = z_0$, and the angle between the tangent line of $\gamma(t)$ at z_0 and the real axis is $\arg \gamma'(0)$. Then $f(z)$ maps $\gamma(t)$ to a smooth curve $\sigma(t) = f(\gamma(t))$ which passes the point $w_0 = f(z_0)$. Thus, we have $\sigma'(t) = f'(\gamma(t))\gamma'(t)$, $\sigma'(0) = f'(z_0)\gamma'(0)$ and the angle between the tangent line of $\sigma(t)$ and the real axis is

$$\arg \sigma'(0) = \arg f'(z_0) + \arg \gamma'(0),$$

or

$$\arg \sigma'(0) - \arg \gamma'(0) = \arg f'(z_0).$$

In words, the difference between the argument of the tangent vector of $\sigma(t)$ at the point w_0 and the argument of the tangent vector of $\gamma(t)$ at the point z_0 is always $\arg f'(z_0)$, which is independent of $\gamma(t)$. For any two smooth curves $\gamma_1(t)$ and $\gamma_2(t)$ $(0 \le t \le 1)$ passing through z_0 with $\gamma_1(0) = \gamma_2(0) = z_0$, their images under the map $f(z)$ are smooth curves $\sigma_1(t)$ and $\sigma_2(t)$ respectively and they pass through the point $w_0 = f(z_0)$. Thus

$$\arg \sigma_2'(0) - \arg \gamma_2'(0) = \arg \sigma_1'(0) - \arg \gamma_1'(0),$$

that is

$$\arg \sigma_2'(0) - \arg \sigma_1'(0) = \arg \gamma_2'(0) - \arg \gamma_1'(0).$$

Hence, the angle between $\gamma_1(t)$ and $\gamma_2(t)$ at the point z_0 is equal to the angle between $\sigma_1(t)$ and $\sigma_2(t)$ at the point $w_0 = f(z_0)$. In other words, at points where the derivative of $f(z)$ is not zero, the magnitude and the direction of the angle between two smooth curves are preserved under the map $w = f(z)$. Since

$$f'(z_0) = \lim_{z \to z_0} \frac{f(z) - f(z_0)}{z - z_0},$$

consider any curve $\gamma(t)$ which passes z_0 and with the image $\sigma(t)$ under the map $f(z)$, we have

$$\lim_{\substack{z \to z_0 \\ z \in \gamma}} \frac{|f(z) - f(z_0)|}{|z - z_0|} = \lim_{\substack{z \to z_0 \\ z \in \gamma}} \frac{|w - w_0|}{|z - z_0|} = |f'(z_0)|.$$

This implies that the limit of the ratio between the distance of image points of z and z_0 and the distance of z and z_0 is independent of the curve. The absolute value $|f'(z_0)|$ is a scaling factor of $f(z)$ at the point z_0.

The above two properties are called *conformality*. And the holomorphic mapping $f(z)$ on D is also called a *conformal mapping* (if $f'(z) \neq 0$). We will discuss it in more details in Chapter 4.

1.4 Complex Integration

If $f(t) = u(t) + iv(t)$ is a complex valued function defined on a real interval $[a, b]$, then

$$\int_a^b f(t)\,dt = \int_a^b u(t)\,dt + i\int_a^b v(t)\,dt.$$

If γ is a piecewise differentiable arc with equation $z = z(t), (a \leq t \leq b)$, $f(z)$ is defined on γ and is continuous, then $f(z(t))$ is also a continuous function of t.

Define

$$\int_\gamma f(z)\,dz = \int_a^b f(z(t))z'(t)\,dt$$

as the integral of $f(z)$ along the curve γ. This integral is independent of changes of perimeters. If $t = t(\tau)$ is a increasing function mapping $\tau\,(\alpha \leq \tau \leq \beta)$ to $t\,(a \leq t \leq b)$, and $t(\tau)$ is piecewise differentiable, then

$$\begin{aligned}\int_a^b f(z(t))z'(t)\,dt &= \int_\alpha^\beta f(z(t(\tau)))z'(t(\tau))\,d\tau \\ &= \int_\alpha^\beta f(z(t(\tau)))\frac{dz(t(\tau))}{d\tau}\,d\tau.\end{aligned}$$

The same result can be obtained if we define the line integral by Riemann sum. Thus, we get some properties similar to integrals of real functions. For instance

$$\int_{-\gamma} f(z)\,dz = -\int_\gamma f(z)\,dz;$$

If $\gamma = \gamma_1 + \gamma_2 + \cdots + \gamma_n$, then

$$\int_{\gamma_1+\gamma_2+\cdots+\gamma_n} f(z)\,dz = \int_{\gamma_1} f(z)\,dz + \int_{\gamma_2} f(z)\,dz + \cdots + \int_{\gamma_n} f(z)\,dz.$$

Since there are not many differences that need to be emphasized, we do not go into too much details here.

Now we focus on the part of complex analysis which corresponds to the third part of calculus that we mentioned before. This is the complex Green's Theorem. In order to write in general notations, we use complex exterior differential forms. Consider z and $\bar{z}$ as independent variables, we define the *exterior product* of differentials as:

$$dz \wedge dz = 0, \quad d\bar{z} \wedge d\bar{z} = 0, \quad dz \wedge d\bar{z} = -d\bar{z} \wedge dz,$$

where

$$dz = dx + idy, \quad d\bar{z} = dx - idy.$$

Then

$$\begin{aligned} d\bar{z} \wedge dz &= (dx - idy) \wedge (dx + idy) = -idy \wedge dx + idx \wedge dy \\ &= 2idx \wedge dy = 2idA, \end{aligned}$$

where dA is the two dimensional area element.

Similar to the real case, we define the *exterior differential form* of degree zero to be function $f(z, \bar{z})$; the exterior differential form of degree one to be $\omega_0 dz + \omega_2 d\bar{z}$, where ω_1 and ω_2 are functions of z and $\bar{z}$; the exterior differential form of degree two to be $\omega_0 dz \wedge d\bar{z}$, where ω_0 is function of z and $\bar{z}$. The action of the exterior differential operator d on the exterior differential form ω is defined as

$$d\omega = \partial\omega \wedge dz + \bar{\partial}\omega \wedge d\bar{z},$$

where $\bar{\partial} = \partial/\partial\bar{z}$, $\partial = \partial/\partial z$. Obviously, $dd\omega = 0$ is also true for any exterior differential form ω. The Green's Theorem in complex form is:

Theorem 1.13 *Suppose that $\omega = \omega_1 dz + \omega_2 d\bar{z}$ is an exterior differential form of degree one on a region Ω, where $\omega_1 = \omega_1(z, \bar{z})$, $\omega_2 = \omega_2(z, \bar{z})$ are differentiable functions of z, $\bar{z}$, and $\partial\Omega$ is the boundary of Ω. Let $\partial = \partial/\partial z$, $\bar{\partial} = \partial/\partial\bar{z}$, and $d = \partial + \bar{\partial}$ be the exterior differential operator. Then*

$$\int_{\partial\Omega} \omega = \iint_{\Omega} d\omega. \tag{1.14}$$

Proof Let $\omega_1 = \xi_1 + i\eta_1$, $\omega_2 = \xi_2 + i\eta_2$, where ξ_1, η_1, ξ_2 and η_2 are real valued functions. Then

$$\begin{aligned} \omega = \omega_1 dz + \omega_2 d\bar{z} &= (\xi_1 + i\eta_1)(dx + idy) + (\xi_2 + i\eta_2)(dx - idy) \\ &= ((\xi_1 + \xi_2)dx + (-\eta_1 + \eta_2)dy) + i((\eta_1 + \eta_2)dx + (\xi_1 - \xi_2)dy). \end{aligned}$$

Also

$$
\begin{aligned}
d\omega &= \partial(\omega_1 dz + \omega_2 d\bar{z}) + \bar{\partial}(\omega_1 dz + \omega_2 d\bar{z}) \\
&= \frac{\partial \omega_1}{\partial z} dz \wedge dz + \frac{\partial \omega_2}{\partial z} dz \wedge d\bar{z} + \frac{\partial \omega_1}{\partial \bar{z}} d\bar{z} \wedge dz + \frac{\partial \omega_2}{\partial \bar{z}} d\bar{z} \wedge d\bar{z} \\
&= \left(\frac{\partial \omega_1}{\partial \bar{z}} - \frac{\partial \omega_2}{\partial z} \right) d\bar{z} \wedge dz \\
&= \left[\frac{1}{2} \left(\frac{\partial}{\partial x} + i \frac{\partial}{\partial y} \right) (\xi_1 + i\eta_1) - \frac{1}{2} \left(\frac{\partial}{\partial x} + i \frac{\partial}{\partial y} \right) (\xi_2 + i\eta_2) \right] 2idA \\
&= \left[- \left(\frac{\partial \xi_1}{\partial y} + \frac{\partial \eta_1}{\partial x} + \frac{\partial \xi_2}{\partial y} - \frac{\partial \eta_2}{\partial x} \right) + i \left(\frac{\partial \xi_1}{\partial x} - \frac{\partial \eta_1}{\partial y} - \frac{\partial \xi_2}{\partial x} - \frac{\partial eta_2}{\partial y} \right) \right] dA
\end{aligned}
$$

By Green's Theorem, we have

$$
\begin{aligned}
&\int_{\partial\Omega} (\xi_1 + \xi_2)\, dx + (-\eta_1 + \eta_2)\, dy \\
&\qquad = \iint_{\Omega} \left(-\frac{\partial}{\partial x}(\eta_1 - \eta_2) - \frac{\partial}{\partial y}(\xi_1 + \xi_2) \right) dA
\end{aligned}
$$

and

$$
\begin{aligned}
&\int_{\partial\Omega} (\eta_1 + \eta_2)\, dx + (\xi_1 - \xi_2)\, dy \\
&\qquad = \iint_{\Omega} \left(\frac{\partial}{\partial x}(\xi_1 - \xi_2) - \frac{\partial}{\partial y}(\eta_1 + \eta_2) \right) dA.
\end{aligned}
$$

Therefore, (1.14) follows.

The equation (1.14) also holds in complex Euclidean spaces of higher dimension and in complex manifolds. This is a special case of the general form. The general form is also refereed to as the Stokes's Theorem. One of the starting points of the next chapter is equation (1.14).

1.5 Elementary Functions

In Calculus, the following three kinds of functions and their compositions are called *elementary functions*:

(1) Power functions, polynomials, rational fractions and their inverses.

(2) Trigonometric functions and their inverses.

(3) Exponential functions and their inverses—logarithmic functions.

Some of these functions are easily defined in the complex field, such as polynomial functions, simply by changing the real variable to a complex variable. For some other functions, such as $\sin z$ and e^z, we need to find out what they represent when the variables are complex. These functions must be redefined. The definitions should have a clear expression and should be the same as the corresponding definitions in calculus when the variables are real. A natural idea of defining them is by the series.

Let y be a real number. Then

$$e^y = 1 + \frac{y}{1!} + \frac{y^2}{2!} + \cdots + \frac{y^n}{n!} + \cdots$$

and obviously

$$\begin{aligned} e^{iy} &= 1 + \frac{iy}{1!} + \frac{(iy)^2}{2!} + \cdots + \frac{(iy)^n}{n!} + \cdots \\ &= 1 + \frac{iy}{1!} - \frac{y^2}{2!} - \frac{iy^3}{3!} + \frac{y^4}{4!} + \cdots \\ &= \left(1 - \frac{y^2}{2!} + \frac{y^4}{4!} - \cdots\right) + i\left(\frac{y}{1!} - \frac{y^3}{3!} + \frac{y^5}{5!} - \cdots\right). \end{aligned}$$

It is well known that

$$\cos y = 1 - \frac{y^2}{2!} + \frac{y^4}{4!} - \cdots,$$

$$\sin y = \frac{y}{1!} - \frac{y^3}{3!} + \frac{y^5}{5!} - \cdots.$$

Therefore, we have

$$e^{iy} = \cos y + i \sin y. \tag{1.15}$$

This is the famous *Euler formula.*

Since for any complex number $z = x + iy$,

$$e^z = e^x(\cos y + i \sin y), \tag{1.16}$$

we can consider (1.15) as a corollary of (1.16). Euler formula is very important, it demonstrated that the exponential function and the trigonometric functions can be expressed by each other. From (1.15), we have

$$\cos y = \frac{e^{iy} + e^{-iy}}{2}, \quad \sin y = \frac{e^{iy} - e^{-iy}}{2i}.$$

So for any complex number $z = x + iy$, we can define

$$\cos z = \frac{e^{iz} + e^{-iz}}{2}, \quad \sin z = \frac{e^{iz} - e^{-iz}}{2i}. \tag{1.17}$$

We can also define $\tan z = \sin z / \cos z$, etc. From (1.17) we get that, if y is a real number, then $\cos iy = \cosh y$ and $\sin iy = i \sinh y$.

By definition (1.16), the inverse function of e^z, denoted by $\operatorname{Log} z$, can be defined as all complex numbers w which satisfy $e^w = z$. And $w = \operatorname{Log} z$ is called the *logarithm* of z. Similarly, we can define $\arccos z$ and $\arcsin z$, the inverse functions of $\cos z$ and $\sin z$.

For the power function z^α, the definition is clear when α is an integer. For any complex number α, we can naturally define

$$w = z^\alpha = e^{\alpha \operatorname{Log} z} \tag{1.18}$$

There will be further discussions about properties of the functions defined by (1.16), (1.17) and (1.18). As we can see, these three kinds of functions seem unrelated to each other in calculus, but they become the same kind of function in the complex number field—the exponential functions and their inverses. Trigonometric functions and their inverses, power functions and their inverses can be expressed by exponential functions and their inverses. The crucial step is the Euler formula. This is a very profound formula. When $y = \pi$, (1.15) becomes $e^{i\pi} = -1$, and this equation connects four important constants in mathematics, e, π, i and -1. Also, the very useful *De Moivre formula*

$$(\cos y + i \sin y)^n = \cos ny + i \sin ny$$

is a corollary of (1.15).

Now, we discuss some properties of the elementary functions defined above.

For the exponential function, we have:

(1) It never vanishes, $e^z \neq 0$, since $|e^z| = e^x > 0$.

(2) For any $z_1 = x_1 + iy_1$ and $z_2 = x_2 + iy_2$, we have $e^{z_1}e^{z_2} = e^{z_1+z_2}$ since

$$\begin{aligned} e^{z_1}e^{z_2} &= e^{x_1}(\cos y_1 + i \sin y_1)e^{x_2}(\cos y_2 + i \sin y_2) \\ &= e^{x_1+x_2}(\cos(y_1 + y_2) + i \sin(y_1 + y_2)) \\ &= e^{z_1+z_2}. \end{aligned}$$

(3) The period of e^z is $2\pi i$, since $e^{2\pi i} = 1$.

(4) The function e^z is holomorphic on $\mathbb{C}$ and $(e^z)' = e^z$. Indeed, since

$$u(x,y) = e^x \cos y, \quad v(x,y) = e^x \sin y,$$

it follows that

$$u_x = v_y = e^x \cos y, \quad u_y = -v_x = -e^x \sin y$$

are continuous functions on $\mathbb{C}$. By Theorem 1.12, e^z is holomorphic on $\mathbb{C}$ and

$$(e^z)' = u_x + iv_x = e^x \cos y + ie^x \sin y = e^z.$$

From the discussion above, we can see that the important properties of e^x in the real number field still hold in the complex number field.

If $w = f(z)$ is a map from a region on z-plane to a region on w-plane, the map is called *univalent* if it is one-to-one.

(5) The regions where e^z is univalent:

Let $z_1 = x_1 + iy_1$ and $z_2 = x_2 + iy_2$ be such that $e^{z_1} = e^{z_2}$. Then

$$e^{x_1}(\cos y_1 + i \sin y_1) = e^{x_2}(\cos y_2 + i \sin y_2)$$

and

$$e^{x_1} e^{iy_1} = e^{x_2} e^{iy_2}.$$

Hence, $x_1 = x_2$, $y_1 = y_2 + 2k\pi$, and $z_1 - z_2 = 2k\pi i$ where k is an integer. Therefore, we can choose the band region $2k\pi < y < 2(k+1)\pi$, $k = 0, \pm 1, \pm 2, \cdots$ as the regions where e^z is univalent. For instance, e^z maps the belt region $z = x + iy$, $0 < y < 2\pi$ one-to-one to the region in the complex plane with the positive part of real axis removed $E = \mathbb{C} \backslash \{z | \, z \geq 0\}$.

Now we look at the inverse of the exponential function, the logarithmic function.

For $z \neq 0$, if a complex number w satisfies $e^w = z$, then w is a *logarithm* of z, and denoted by $\operatorname{Log} z$. Since the exponential function is a periodical function, $\operatorname{Log} z$ is an (infinite) multiple valued function.

Let

$$z = re^{i\theta}, \quad w = u + iv.$$

Then

$$e^{u+iv} = re^{i\theta}.$$

It follows that $e^u = r$, $v = \theta + 2k\pi$, where k is an integer. Thus

$$w = \log r + (\theta + 2k\pi)i$$

or

$$w = \log|z| + i\,\mathrm{Arg}\, z,$$

where $\mathrm{Arg}\, z = \theta + 2k\pi$ represent the arguments of z for different ks.

Since $\mathrm{Arg}(z_1 z_2) = \mathrm{Arg}\, z_1 + \mathrm{Arg}\, z_2$, the logarithm has the property

$$\mathrm{Log}(z_1 z_2) = \mathrm{Log}\, z_1 + \mathrm{Log}\, z_2.$$

From the discussion of the exponential function, we can see that, if we choose the principle value of the argument of z between 0 and 2π on the region $D = \mathbb{C} \setminus \{z|\, z \geq 0\}$, then each of the function

$$w_k(z) = \log|z| + i(\arg z + 2k\pi) \quad (k = 0, \pm 1, \pm 2, \cdots)$$

maps D one-to-one to the band region $E_k = \{v|\, 2k\pi < v < 2(k+1)\pi\}$ parallel to the real axis. They are the inverses of the exponential function $z = e^w$ and are holomorphic on D. Furthermore, $w_k' = 1/z$. The function $w_0(z) = \log|z| + i \arg z$ is called the *principle branch* of $\mathrm{Log}\, z$ and is denoted by $\log z$. That is

$$\log z = \log|z| + i \arg z.$$

Sometimes, we choose $-\pi < \arg z < \pi$ for convenience.

For trigonometric functions, we can directly get the following from the definition (1.17):

(1) Functions $\cos z$, $\sin z$ are holomorphic on $\mathbb{C}$ and

$$(\sin z)' = \cos z, \quad (\cos z)' = -\sin z.$$

(2) Functions $\cos z$, $\sin z$ have period 2π,

$$\sin(z + 2\pi) = \sin z, \quad \cos(z + 2\pi) = \cos z.$$

(3) The function $\cos z$ is even and the function $\sin z$ is odd,

$$\sin(-z) = -\sin z, \quad \cos(-z) = \cos z.$$

(4) The sum and difference identities:

$$\sin(z_1 \pm z_2) = \sin z_1 \cos z_2 \pm \cos z_1 \sin z_2,$$

$$\cos(z_1 \pm z_2) = \cos z_1 \cos z_2 \mp \sin z_1 \sin z_2.$$

(5) The fundamental identities between $\sin z$ and $\cos z$:

$$\sin^2 z + \cos^2 z = 1, \quad \sin\left(\frac{\pi}{2} - z\right) = \cos z.$$

(6) The zeros of $\sin z$ are $z = k\pi$ and the zeros of $\cos z$ are $z = \pi/2 + k\pi$, $k = 0, \pm 1, \pm 2, \cdots$.

As we can see from the above that the main properties of $\sin x$ and $\cos x$ on the real number field still hold on the complex number field. There are some different properties of $\cos z$ and $\sin z$ for complex z.

(7) The functions $|\sin z|$ and $|\cos z|$ are unbounded. Indeed, by (4), we have

$$\begin{aligned}|\sin z|^2 &= |\sin(x+iy)|^2 = |\sin x \cos iy + \cos x \sin iy|^2 \\ &= |\sin x \cosh y + i \cos x \sinh y|^2 = \sinh^2 y + \sin^2 x.\end{aligned}$$

This is an unbounded function. Similarly, $|\cos z|^2 = \cosh^2 y - \sin^2 x$ is also an unbounded function.

(8) The region where $\sin z$ and $\cos z$ are univalent:

Consider

$$w = \cos z = \frac{e^{iz} + e^{-iz}}{2},$$

this is a composition of the functions $\xi = iz$, $\zeta = e^{\xi}$ and $w = (1/2)(\zeta + 1/\zeta)$. The first function is a rotation, it is univalent everywhere. The necessary and sufficient condition for the second function to be univalent is that no two distinct points ξ_1 and ξ_2 in the region satisfy $\xi_2 - \xi_1 = 2k\pi i$, where k is an integer. On the z-plane, this condition becomes that no two distinct points z_1 and z_2 satisfy $z_2 - z_1 = 2k\pi$. The necessary and sufficient condition for the third function to be univalent is that no two distinct points ζ_1 and ζ_2 in the region satisfy $\zeta_1\zeta_2 = 1$. On the z-plane, this condition becomes that no two distinct points z_1 and z_2 satisfy $e^{iz_1}e^{iz_2} = 1$. And this is equivalent to $z_1 + z_2 = 2k\pi$. Thus, the band region $0 < \operatorname{Re} z < \pi$ is one of the regions where $\cos z$ is univalent.

The function $\xi = iz$ maps $0 < \operatorname{Re} z < \pi$ one-to-one to $0 < \operatorname{Im} z < \pi$, the function $\zeta = e^{\xi}$ maps $0 < \operatorname{Im} z < \pi$ one-to-one to the upper half plane, $\operatorname{Im} \zeta > 0$, and finally the function $w = (1/2)(\zeta + 1/\zeta)$ maps the upper half plane to the w-plane with $-\infty < u \le -1$ and $1 \le u < +\infty$ removed. In other words, $w = \cos z$ maps $0 < \operatorname{Re} z < \pi$ one-to-one to the w-plane with $-\infty < u < -1$, $v = 0$ and $1 \le u < \infty$, $v = 0$ removed.

Similar discussions can be made for $w = \sin z$ and $w = \tan z$.

For the inverses of trigonometric functions, consider $w = \arccos z$, that is, $\cos w = z$.

Since

$$\cos w = \frac{1}{2}(e^{iw} + e^{-iw}) = z$$

is a quadratic equation with respect to e^{iw}, and has the roots $e^{iw} = z \pm \sqrt{z^2 - 1}$. It follows that

$$\begin{aligned} w = \arccos z &= -i \operatorname{Log}(z \pm \sqrt{z^2 - 1}) \\ &= \pm i \operatorname{Log}(z + \sqrt{z^2 - 1}). \end{aligned}$$

Therefore, $\arccos z$ is a multiple-valued function which has infinitely many values. This implies the periodicity of $\cos w$. The function $\arcsin z$ can be defined as $\pi/2 - \arccos z$ and a similar discussion follows.

As the last one, we look at the power function.

By the definition (1.18), if $\alpha = a + ib$, then

$$\begin{aligned} z^\alpha &= e^{\alpha \operatorname{Log} z} = e^{(a+ib)(\log|z| + i(\arg z + 2k\pi))} \\ &= e^{a \log|z| - b(\arg z + 2k\pi)} \cdot e^{i(b \log|z| + a(\arg z + 2k\pi))}, \end{aligned}$$

where k is an integer. Let

$$\rho_k = e^{a \log|z| - b(arg z + 2k\pi)},$$

$$\theta_k = b \log|z| + a(arg z + 2k\pi).$$

Then

$$w = z^\alpha = \rho_k e^{i\theta_k}, \quad |w| = \rho_k.$$

Hence, if $b \neq 0$, then $w = z^\alpha$ is a function with infinite many values; if $b = 0$, then α is the real number a, and

$$w = z^\alpha = e^{a \log|z|} e^{i(\arg z + 2k\pi)a} = |z|^a e^{a(\arg z + 2k\pi)i}.$$

In the later case, all of the values of z^α are on the circle $|w| = |z|^a$. Hence, we have that

(1) If $\alpha = a = n$, where n is an integer, then $z^\alpha = z^n$ is single valued;

(2) If $\alpha = a = p/q$, where p/q is a reduced fraction with $0 < p < q$, then

$$z^\alpha = e^{\frac{p}{q} \log|z|} e^{i\frac{p}{q}(\arg z + 2k\pi)} = |z|^{\frac{p}{q}} e^{i\frac{p}{q} \arg z} e^{i\frac{p}{q} 2k\pi}.$$

When $k = 0, 1, 2, \cdots, q-1$, the corresponding q different values of $(p/q)2k\pi$ are not congruent modulo 2π, but $(p/q)2k\pi$ and one of these q values will be congruent modulo 2π if k is an integer other than $0, 1, 2, \cdots, q-1$. Hence z^α only has these q different values for any given z.

(3) If $\alpha = a$ is an irrational number, then z^α has infinite many values.

1.6 Complex Series

A part of the theory about series in calculus also can be generalized to the complex number field without any difficulties.

A sequence of functions $\{f_n(z)\}$ converges uniformly to $f(z)$ on a set $E \subset \mathbb{C}$ if for any $\varepsilon > 0$, there exists a positive integer n_0 such that

$$|f_n(z) - f(z)| < \varepsilon$$

for all $n \geq n_0$ and all $z \in E$.

Similar to calculus, we can prove that, the limit function of a uniformly convergent sequence of continuous functions is also continuous.

Cauchy Criterion for Convergence A sequence of functions $\{f_n(z)\}$ converges uniformly on a set $E \subset \mathbb{C}$ if and only if for any $\varepsilon > 0$, there exists a positive integer n_0 such that

$$|f_m - f_n| < \varepsilon$$

for all $m, n \geq n_0$ and all $z \in E$.

Weierstrass M-test Suppose that

$$f_1(z) + f_2(z) + \cdots + f_n(z) + \cdots$$

is a series of functions defined on a set $E \subset \mathbb{C}$, and $a_1 + a_2 + a_3 + \cdots$ is a series of positive numbers. If there exist a positive integer n_0 and a constant $M > 0$ such that $f_n(z) \leq Ma_n$ for $n \geq n_0$ and all $z \in E$, then the series $\sum_{n=1}^{\infty} f_n(z)$ converges uniformly on E if the series $\sum_{n=1}^{\infty} a_n$ converges.

Especially, for power series

$$a_0 + a_1 z + a_2 z^2 + \cdots + a_n z^n + \cdots, \tag{1.19}$$

we have

Theorem 1.14 *(**Abel Theorem**) For the power series (1.19), there exists a number R, $0 \leq R \leq \infty$, called the radius of convergence, with the following properties:*

(1) For any z satisfying $|z| < R$, the series is absolutely convergent. The series is uniformly convergent on $|z| \leq \rho$, where $0 \leq \rho < R$.

(2) If z satisfies $|z| > R$, then the terms of the series are unbounded and the series is divergent.

(3) The sum of the series is a holomorphic function in the disc $|z| < R$. The derivative of the sum can be obtained by term-wise differentiation and the resulting series has the same radius of convergence.

The closed disc $|z| \leq R$ is called the *disc of convergence*. The series is not necessarily convergent or divergent on the boundary of this disc. The radius of convergence R can be determined by *Hadamard's formula*

$$R = \frac{1}{\limsup_{n\to\infty} \sqrt[n]{|a_n|}}. \tag{1.20}$$

If $R = 0$, then the series (1.19) is divergent except at $z = 0$;

If $R = \infty$, then the series (1.19) is convergent everywhere.

Proof of Theorem 1.14 If $|z| < R$, we can find a ρ with $|z| < \rho < R$. It follows that $1/\rho > 1/R$. By (1.20), there exists an n_0 such that $|a_n|^{\frac{1}{n}} < 1/\rho$ or $|a_n| < 1/\rho^n$ for all $n \geq n_0$. Therefore, $|a_n z^n| < (|z|/\rho)^n$ for $n \geq n_0$. Since $\sum_{n=0}^{\infty}(|z|/\rho)^n$ is convergent for $|z| < \rho$, by the Weierstrass M-test, $\sum_{n=0}^{\infty} a_n z^n$ is absolutely convergent. In order to show that the series is uniformly convergent when $|z| \leq \rho(< R)$, we choose a ρ' and an n_1 such that $\rho < \rho' < R$ and $|a_n z^n| \leq (\rho/\rho')^n$, for $n \geq n_1$. The series (1.19) is uniformly convergent on $|z| \leq \rho$ by Weierstrass M-test.

If $|z| > R$, we can find a ρ with $R < \rho < |z|$. Since $1/\rho < 1/R$, there exists an n_2 and a subsequence $\{m_i\}$ such that $|a_{m_i}|^{1/m_i} > 1/\rho$ or $|a_{m_i}| > 1/\rho^{m_i}$ for $m_i \geq n_2$. Therefore, $|a_n z^n| > (|z|/\rho)^n$ for infinitely many n, and the terms of the series are unbounded.

The series $\sum_{n=1}^{\infty} n a_n z^n$ and $\sum_{n=1}^{\infty} a_n z^n$ have the same radius of convergence because of $\lim_{n\to\infty} \sqrt[n]{n} = 1$.

For $|z| < R$, suppose

$$f(z) = \sum_{n=0}^{\infty} a_n z^n = S_n(z) + R_n(z),$$

where

$$S_n(z) = a_0 + a_1 z + \cdots + a_{n-1} z^{n-1}, \quad R_n(z) = \sum_{k=n}^{\infty} a_k z^k.$$

Let

$$f_1(z) = \sum_{n=1}^{\infty} n a_n z^{n-1} = \lim_{n\to\infty} S'_n(z).$$

We need to show that $f_1(z) = f'(z)$. Choose a ρ and a point z_0 such that $0 < \rho < R$ and $|z_0| < \rho$. Then

$$\frac{f(z)-f(z_0)}{z-z_0} - f_1(z)$$
$$= \left(\frac{S_n(z)-S_n(z_0)}{z-z_0} - S'_n(z_0)\right) + (S'_n(z_0) - f_1(z)) + \left(\frac{R_n(z)-R_n(z_0)}{z-z_0}\right).$$

The last term on the right hand side of this equation can be written as

$$\sum_{k=n}^{\infty} a_k(z^{k-1} + z^{k-2}z_0 + \cdots + z z_0^{k-2} + z_0^{k-1}),$$

if $z \neq z_0$ and $|z| < \rho < R$. Thus

$$\left|\frac{R_n(z)-R_n(z_0)}{z-z_0}\right| < \sum_{k=n}^{\infty} k|a_k|\rho^{k-1}.$$

This is a remainder of a convergent series, so there exists an n_3 such that

$$\left|\frac{R_n(z)-R_n(z_0)}{z-z_0}\right| < \frac{\varepsilon}{3}$$

for $n \geq n_3$.

Since $f_1(z) = \lim_{n\to\infty} S'_n(z)$, there exists an n_4 such that $|S'_n(z_0) - f_1(z)| < \varepsilon/3$, for $n \geq n_4$. For a fixed n with $n > n_3$ and $n > n_4$, we can find a $\delta > 0$ by the definition of derivative such that

$$\left|\frac{S_n(z)-S_n(z_0)}{z-z_0} - S'_n(z_0)\right| < \frac{\varepsilon}{3}$$

for $0 < |z - z_0| < \delta$. Therefore, we have

$$\left|\frac{f(z)-f(z_0)}{z-z_0} - f_1(z_0)\right| < \varepsilon$$

for $0 < |z - z_0| < \delta$. Hence $f'(z_0) = f_1(z_0)$ and we complete the proof.

Repeat the process in the proof of the last theorem, we have

$$f^k(z) = k!a_k + \frac{(k+1)!}{1!}a_{k+1}z + \frac{(k+2)!}{2!}a_{k+2}z^2 + \cdots$$

for any positive integer k. Hence, $a_k = f^k(0)/k!$ and the power series can be written as

$$f(z) = f(0) + f'(0) + \frac{f"(0)}{2!}z^2 + \cdots + \frac{f^{(n)}(0)}{n!}z^n + \cdots$$

This is known as *Taylor-Maclaurin series*. We obtained this representation under the assumption that the power series expansion of $f(z)$ exists and is unique. We will prove the assertion that "*Every holomorphic function has a Taylor expansion*" in the next chapter.

From property (4) of the exponential function, $(e^z)' = e^z$, we get the Taylor series for e^z

$$e^z = 1 + \frac{z}{1!} + \frac{z^2}{2!} + \cdots + \frac{z^n}{n!} \cdots .$$

By definition (1.17), the Taylor series for $\cos z$ and $\sin z$ are

$$\cos z + 1 - \frac{z^2}{2!} + \frac{z^4}{4!} - \frac{z^6}{6!} + \cdots ,$$

$$\sin z = z - \frac{z^3}{3!} + \frac{z^5}{5!} - \frac{z^7}{7!} + \cdots .$$

These three series are convergent on the entire plane by Hadamard's formula.

Finally we state two results about function series.

Theorem 1.15 *(1) If $\{f_n(z)\}$ is continuous on a set A, and $\sum_{n=1}^{\infty} f_n(z)$ converges uniformly to $f(z)$ on A, then $f(z)$ is continuous on A.*

(2) If $\{f_n(z)\}$ is continuous on a rectifiable curve γ, and $\sum_{n=1}^{\infty} f_n(z)$ converges uniformly to $f(z)$, then

$$\int_\gamma f(z)\,dz = \sum_{n=1}^{\infty} \int_\gamma f_n(z)\,dz.$$

We omit the proofs here since they are the same as in calculus.

Exercise I

1. Verify the Newton-Leibniz formula, the Green's theorem, the Stokes's theorem and the Gauss's theorem by (1.9).

2. (1) Find the modulus and the principle values of the arguments for the following complex numbers:

(i) $2i$; (ii) $1-i$; (iii) $3+4i$; (iv) $-5+12i$.

(2) Express the following complex numbers in the form of $x+iy$ where x and y are real numbers.

(i) $(1+3i)^3$; (ii) $\dfrac{10}{4-3i}$; (iii) $\dfrac{2-3i}{4+i}$;

(iv) $(1+i)^n+(1-i)^n$, where n is a positive integer.

(3) Find the absolute values (modulus) of the following complex numbers:

(i) $-3i(2-i)(3+2i)(1+i)$; (ii) $\dfrac{(4-3i)(2-i)}{(1+i)(1+3i)}$.

3. Show that

$$\frac{\pi}{4}=4\arctan\frac{1}{5}-\arctan\frac{1}{239}$$

by calculating $(5-i)^4(1+i)$.

4. Let $z=x+iy$, where x and y are real numbers. Find the real part and the imaginary part of each of the following complex numbers:

(1) $\dfrac{1}{\bar{z}}$; (2) z^2; (3) $\dfrac{1+z}{1-z}$; (4) $\dfrac{z}{z^2+1}$.

5. Solve the quadratic equation

$$z^2+(\alpha+i\beta)z+\gamma+i\delta=0,$$

where α, β, γ and δ are real numbers.

6. Assume $|z|=r>0$. Show that:

$$\operatorname{Re} z=\frac{1}{2}\left(z+\frac{r^2}{z}\right),\quad \operatorname{Im} z=\frac{1}{2i}\left(z-\frac{r^2}{z}\right).$$

7. Show that:

(1) If $|a|<1$, $|b|<1$, then

$$\left|\frac{a-b}{1-\bar{a}b}\right|<1;$$

(2) If $|a|=1$ or $|b|=1$, then

$$\left|\frac{a-b}{1-\bar{a}b}\right|=1.$$

Discussing the case $|a|=1$ and $|b|=1$.

8. Prove the Lagrange equation in complex form

$$\left|\sum_{i=1}^{n}a_ib_i\right|^2=\sum_{i=1}^{n}|a_i|^2\sum_{i=1}^{n}|b_i|^2-\sum_{1\le i<j\le n}|a_i\bar{b}_j-a_j\bar{b}_i|^2.$$

Derive the Cauchy inequality

$$\left|\sum_{i=1}^{n} a_i b_i\right|^2 \leq \sum_{i=1}^{n} |a_i|^2 \sum_{i=1}^{n} |b_i|^2$$

from the Lagrange equation. The equality holds if and only if a_k and $\bar{b}_k$ are proportional, where $k = 1, \cdots, n$.

9. Show that a_1, a_2, a_3 are the vertices of a equilateral triangle if and only if

$$a_1^2 + a_2^2 + a_3^2 = a_1 a_2 + a_2 a_3 + a_3 a_1.$$

10. Show that the complex numbers α, β, γ are collinear if and only if

$$\begin{vmatrix} \alpha & \bar{\alpha} & 1 \\ \beta & \bar{\beta} & 1 \\ \gamma & \bar{\gamma} & 1 \end{vmatrix} = 0.$$

11. Find the corresponding points of $1-i$, $4+3i$ on the Riemann sphere S^2.

12. Two points z_1, z_2 on the $\mathbb{C}$-plane correspond to the two end points of a diagonal of the Riemann sphere S^2 if and only if $z_1 z_2 = -1$.

13. Show that the equation of a circle is

$$A|z|^2 + Bz + \bar{B}\bar{z} + C = 0$$

where A, C are real numbers and $|B|^2 > AC$. Also show that the image of this circle on the Riemann sphere S^2 is a great circle if and only if $A+C = 0$.

14. Let $d(z_1, z_2)$ be the spherical distance of z_1 and z_2 in $\mathbb{C}$, that is, the spherical distance between the corresponding points Z_1 and Z_2 on the Riemann sphere S^2. Prove

$$d(z_1, z_2) = \frac{2|z_1 - z_2|}{\sqrt{(1 + |z_1|^2)(1 + |z_2|^2)}},$$

and

$$d(z_1, \infty) = \frac{2}{\sqrt{1 + |z_1|^2}}.$$

15. Explain the geometric meaning of the following relations.

(i) $\left|\dfrac{z - z_1}{z - z_2}\right| \leq 1$, $z_1 \neq z_2$; (ii) $\operatorname{Re} \dfrac{z - z_1}{z - z_2} = 0$, $z_1 \neq z_2$;

(iii) $0 < \arg \dfrac{z + i}{z - i} < \dfrac{\pi}{4}$;

(iv) $|z+c|+|z-c|\le 2a$, $a>0$, $|c|<a$.

16. Prove Heine-Borel theorem and Bolzano-Weierstrass theorem on the complex plane.

17. Show that the sequence $\{z_n\}$ converges to z_0 if and only if the sequences $\{\operatorname{Re} z_n\}$, $\{\operatorname{Im} z_n\}$ converges to $\operatorname{Re} z_0$, $\operatorname{Im} z_0$ respectively.

18. Show that

(1) If $\lim_{n\to\infty} z_n = a$, $\lim_{n\to\infty} z'_n = b$, then

$$\lim_{n\to\infty}\frac{1}{n}\sum_{k=1}^{n} z_k z'_{n-k} = ab.$$

(2) If $\lim_{n\to\infty} = A$, then

$$\lim_{n\to\infty}\frac{1}{n}(z_1+z_2+\cdots+z_n) = A$$

(using the result of (1)).

19. Discuss the differentiability of the following functions.

(i) $f(z)=|z|$; (ii) $f(z)=\bar{z}$; (iii) $f(z)=\operatorname{Re} z$.

20. Suppose $g(w)$ and $f(z)$ are holomorphic functions. Show that $g(f(z))$ is also holomorphic.

21. On a region D, show that

(1) If a function $f(z)$ is holomorphic, and $f'(z)$ is identically zero, then $f(z)$ is a constant function.

(2) If $f(z)$ is holomorphic and $f'(z)$ satisfies one of the following conditions:

(i) $\operatorname{Re} f(z)$ is a constant function;

(ii) $\operatorname{Im} f(z)$ is a constant function;

(iii) $|f(z)|$ is a constant function;

(iv) $\arg f(z)$ is a constant function,

then $f(z)$ is a constant function.

22. If $f(z)=u+iv$ is holomorphic and $f'(z)\neq 0$, then the curves $u(x,y)=c_1$ and $v(x,y)=c_2$ are orthogonal, where c_1, c_2 are constants.

23. Show that

(1) If

$$f(z)=u(r,\theta)+iv(r,\theta),\quad z=r(\cos\theta+i\sin\theta),$$

then the C-R equations become

$$u_r=\frac{1}{r}v_\theta,\quad v_r=-\frac{1}{r}u_\theta,$$

and

$$f'(z) = \frac{r}{z}(u_r + iv_r).$$

(2) If $f(z) = R(\cos\varphi + i\sin\varphi)$, then the C-R equations are

$$\frac{\partial R}{\partial r} = -\frac{R}{r}\frac{\partial \varphi}{\partial \theta}, \quad \frac{\partial R}{\partial \theta} = -Rr\frac{\partial \varphi}{\partial r}.$$

24. Suppose that the function $f(z) = u(x,y) + iv(x,y), (z = x + iy)$ is holomorphic on a region D. Show that the Jacobi determinant of u and v with respect to x and y is

$$J = \begin{vmatrix} u_x & u_y \\ v_x & v_y \end{vmatrix} = |f'(z)|^2,$$

and give the geometric meaning of J.

25. Compute:

(1) The integral

$$\int_\gamma x\,dz$$

where $z = x + iy$ and γ is a directed segment from 0 to $1 + i$.

(2) The integral

$$\int_\gamma |z-1|\,|dz|$$

where $\gamma(t) = e^{it},\ 0 \le t \le 2\pi$.

(3) The integral

$$\int_\gamma \frac{1}{z-a}\,dz$$

where $\gamma(t) = a + Re^{it},\ 0 \le t \le 2\pi$ and a is a complex constant.

26. (1) Find the real parts and the imaginary parts of $\cos(x + iy)$, $\sin(x + iy)$ where x, y are real numbers.

(2) Show that

$$\sin iz = i\sinh z, \quad \cos iz = \cosh z, \quad (\sin z)' = \cos z,$$

$$\cos(z_1 + z_2) = \cos z_1 \cos z_2 - \sin z_1 \sin z_2.$$

27. Evaluate

$$\sin i, \quad \cos(2+i), \quad \tan(1+i), \quad 2^i, \quad i^i, \quad (-1)^{2i},$$

$$\log(2-3i), \quad \arccos\frac{1}{4}(3+i).$$

28. (1) Find the values of e^z for $z = \pi i/2, -(2/3)\pi i$.

(2) Find the values of z, if z satisfies $e^z = i$.

29. Find the real part and the imaginary part of z^z, where $z = x + iy$.

30. Show that the roots of $z^n = a$ are the vertices of a regular polygon.

31. Show that the Cauchy criterion for convergence and the Weierstrass M-test hold for complex field.

32. Find the radius of convergence of $\sum_{n=1}^{\infty} a_n z^n$ if

(i) $a_n = n^{1/n}$; (ii) $a_n = n^{\ln n}$;

(iii) $a_n = n!/n^n$; (iv) $a^n = n^n$.

33. Prove Theorem 1.15.

34. Let $f(z)$ be holomorphic on $\mathbb{C}$ and $f(0) = 1$. Show that:

(1) If $f'(z) = f(z)$ for all $z \in \mathbb{C}$, then $f(z) = e^z$.

(2) If for every $z \in \mathbb{C}$ and every $w \in \mathbb{C}$, $f(z+w) = f(z)f(w)$ and $f'(0) = 1$, then $f(z) = e^z$.

35. Let f be a holomorphic function on $\mathbb{C} \setminus (-\infty, 0]$. Show that:

(1) If $f(z) = e^{-f(z)}$ for every $z \in \mathbb{C} \setminus (-\infty, 0]$, then $f(z) = \log z$.

(2) If $f(zw) = f(z) + f(w)$ for every $z \in \mathbb{C} \setminus (-\infty, 0]$, $w \in (0, +\infty)$ and $f'(1) = 1$, then $f(z) = \log z$.

36. If

$$\varphi(z) = \frac{1}{2}\left(z + \frac{1}{z}\right),$$

then φ is univalent on the following four regions:

(1) upper half plane $\{z \in \mathbb{C} : \operatorname{Im} z > 0\}$.

(2) lower half plane $\{z \in \mathbb{C} : \operatorname{Im} z < 0\}$.

(3) unit disc $D(0,1)$ except the center point $\{0\}$, $D(0,1) \setminus \{0\}$.

(4) outside of the unit disc $\{z \in \mathbb{C} : |z| > 1\}$.

37. Find the image of the four regions in Exercise 36 under the map

$$\varphi(z) = \frac{1}{2}\left(z + \frac{1}{z}\right).$$

38. Show that $\cos z$ and $\sin z$ are univalent in the following three regions.

(1) band region $\{z \in \mathbb{C} : \theta_0 < \operatorname{Re} z < \theta_0 + \pi\}$.

(2) half band region $\{z \in \mathbb{C} : \theta_0 < \operatorname{Re} z < \theta_0 + 2\pi, \operatorname{Im} z > 0\}$.

(3) half band region $\{z \in \mathbb{C} : \theta_0 < \operatorname{Re} z < \theta_0 + 2\pi, \operatorname{Im} z < 0\}$, where θ_0 is a fixed real number.

39. Show that $w = \cos z$ maps the half band region $\{z \in \mathbb{C} : 0 < \operatorname{Re} z < 2\pi, \operatorname{Im} z > 0\}$ one-to-one to $\mathbb{C} \setminus [-1, +\infty)$.

40. Show that $w = \sin z$ maps the half band region

$$\{z \in \mathbb{C} : -\frac{\pi}{2} < \operatorname{Re} z < \frac{\pi}{2}, \operatorname{Im} z > 0\}$$

one-to-one to the upper half plane.

41. Show that $f(z) = z/(1-z)^2$ is univalent in $D(0,1)$ and find $f(D(0,1))$.

42. Evaluate the increments of argument of the following functions after z travels along the circle $\{z \in \mathbb{C} : |z| = 2\}$ once counterclockwise.

(i) $(z-1)^{1/2}$; (ii) $(1+z^4)^{1/3}$; (iii) $(z^2+2z-3)^{1/4}$;

(iv) $\left(\dfrac{z-1}{z+1}\right)^{1/2}$; (v) $\left(\dfrac{z^2-1}{z^2+5}\right)^{1/7}$.

43. If $f(z)$ is a univalent holomorphic function on a region U, then the area of $f(U)$ is

$$\iint_U |f'(z)|^2 \, dxdy,$$

where $z = x + iy$.

44. Show that if $\{a_n\}$ and $\{b_n\}$ satisfy the conditions:

(1) the sequence $\{S_n\}$ is bounded, where

$$S_n = \sum_{k=1}^{n} a_k,$$

(2)

$$\lim_{n\to\infty} b_n = 0,$$

(3)

$$\sum_{n=1}^{\infty} |b_n - b_{n+1}| < \infty,$$

then the series

$$\sum_{n=1}^{\infty} a_n b_n$$

is convergent. Moreover, verify that this is a generalization of the Dirichlet criterion and the Abel criterion for series in the real number field.

45. Let

$$\sum_{n=1}^{\infty} a_n$$

be a complex series, and $\overline{\lim}_{n\to\infty} \sqrt[n]{|a_n|} = q$. Prove:

(1) If $q < 1$, then

$$\sum_{n=1}^{\infty} a_n$$

is absolutely convergent.

(2) If $q > 1$, then

$$\sum_{n=1}^{\infty} a_n$$

is divergent.

46. Show that if $a_n \in \mathbb{C} \setminus \{0\}, n = 1, 2, \cdots$, and

$$\overline{\lim_{n\to\infty}} \left| \frac{a_{n+1}}{a_n} \right| = q,$$

then

$$\sum_{n=1}^{\infty} a_n$$

is absolutely convergent if $q < 1$. Discuss the convergence and divergence of the series if $q > 1$.

47. Let $a_n \in \mathbb{C} \setminus \{0\}, (n = 1, 2, \cdots)$ and

$$\overline{\lim_{n\to\infty}} \left| \frac{a_{n+1}}{a_n} \right| = 1.$$

Show that if

$$\overline{\lim_{n\to\infty}}\, n \left(\left| \frac{a_{n+1}}{a_n} \right| - 1 \right) < -1,$$

then

$$\sum_{n=1}^{\infty} a_n$$

is absolutely convergent (*Reabe criterion*).

48. Let $\{a_n\}$ be a series of positive numbers that converges to zero monotonically. Prove:

(1) If R is the radius of convergence of the series

$$\sum_{n=0}^{\infty} a_n z^n,$$

then $R \geq 1$.

(2) The series

$$\sum_{n=0}^{\infty} a_n z^n$$

is convergent on $\partial D(0, R) \setminus \{R\}$, where $\partial D(0, R)$ is the boundary of the disc $D(0, R)$.

49. Show that the power series

$$\sum_{n=0}^{\infty} a_n (z - z_0)^n$$

is uniformly convergent on its disc of convergence D if and only if it is uniformly convergent on $\bar{D}$. Here $z_0 \in \mathbb{C}$ is a fixed point and $\bar{D}$ is the closure of D.

50. Suppose that the radius of convergence of the series

$$\sum_{n=0}^{\infty} a_n z^n = f(z)$$

is 1, and $z_0 \in \partial D(0, 1)$. Prove that if $\lim_{n\to\infty} na_n = 0$ and $\lim_{r\to 1} f(rz_0)$ exists, then

$$\sum_{n=0}^{\infty} a_n z_0^n$$

converges to $\lim_{n\to\infty} f(rz_0)$.

Chapter 2

Cauchy Integral Theorem and Cauchy Integral Formula

2.1 Cauchy-Green Formula (Pompeiu Formula)

Cauchy integral theory is one of the three main theories in the complex analysis of one variable. The theory of analytic functions of one complex variable becomes an independent subject because of the discovery of Cauchy integral theory. A series of results which are different from calculus can be derived by using the Cauchy integral theory.

We start from the Cauchy-Green formula. This is a corollary of Theorem 1.13 (Green's Theorem in complex form) in the last chapter.

Theorem 2.1 ***(Cauchy-Green Formula, Pompeiu Formula)*** *Let $U \subseteq \mathbb{C}$ be a region bounded by a smooth curve and $f(z) = u(x,y)+iv(x,y) \in C^1(\bar{U})$, that is, $u(x,y)$ and $v(x,y)$ have continuous first partial derivative on $\bar{U}$. Then*

$$
\begin{aligned}
f(z) &= \frac{1}{2\pi i}\int_{\partial U} \frac{f(\zeta)}{\zeta - z} d\zeta - \frac{1}{2\pi i}\iint_U \frac{\partial f(\zeta)}{\partial \bar{\zeta}} \cdot \frac{d\bar{\zeta}\wedge d\zeta}{\zeta - z} \\
&= \frac{1}{2\pi i}\int_{\partial U} \frac{f(\zeta)}{\zeta - z} d\zeta - \frac{1}{\pi}\iint_U \frac{\partial f(\zeta)}{\partial \bar{\zeta}} \cdot \frac{dA}{\zeta - z}.
\end{aligned} \tag{2.1}
$$

Proof Let $D(z,\varepsilon) \subseteq U$ be a disc with radius ε and center z. Consider the differential form

$$
\frac{f(\zeta)d\zeta}{\zeta - z}
$$

in $U \setminus D(z,\varepsilon)$ (denoted by $U_{z,\varepsilon}$). Then by Theorem 1.13, we have

$$
\int_{\partial U} \frac{f(\zeta)d\zeta}{\zeta - z} - \int_{\partial D(z,\varepsilon)} \frac{f(\zeta)d\zeta}{\zeta - z} = \iint_{U_{z,\varepsilon}} d_\zeta\left(\frac{f(\zeta)d\zeta}{\zeta - z}\right).
$$

By the definition of d_ζ,

$$\iint_{U_{z,\varepsilon}} d_\zeta\left(\frac{f(\zeta)d\zeta}{\zeta - z}\right) = \iint_{U z,\zeta} (\partial + \bar{\partial})\left(\frac{f(\zeta)d\zeta}{\zeta - z}\right)$$
$$= \iint_{U z,\varepsilon} \partial\left(\frac{f(\zeta)d\zeta}{\zeta - z}\right) + \iint_{U z,\varepsilon} \bar{\partial}\left(\frac{f(\zeta)d\zeta}{\zeta - z}\right).$$

Since

$$\partial\left(\frac{f(\zeta)d\zeta}{\zeta - z}\right) = \frac{\partial}{\partial\zeta}\left(\frac{f(\zeta)}{\zeta - z}\right) d\zeta \wedge d\zeta = 0,$$

$$\bar{\partial}\left(\frac{f(\zeta)d\zeta}{\zeta - z}\right) = \frac{\partial f}{\partial\bar{\zeta}} \frac{d\bar{\zeta} \wedge d\zeta}{\zeta - z} + f\frac{\partial}{\partial\bar{\zeta}}\left(\frac{1}{\zeta - z}\right) d\bar{\zeta} \wedge d\zeta,$$

and because of $\frac{\partial}{\partial\bar{\zeta}}\frac{1}{\zeta - z} = 0$ we have

$$\bar{\partial}\left(\frac{f(\zeta)d\zeta}{\zeta - z}\right) = \frac{\partial f}{\partial\bar{\zeta}} \cdot \frac{d\bar{\zeta} \wedge d\zeta}{\zeta - z},$$

hence

$$\iint_{U_{z,\varepsilon}} d_\zeta\left(\frac{f(\zeta)d\zeta}{\zeta - z}\right) = \iint_{U_{z,\varepsilon}} \frac{\partial f}{\partial\bar{\zeta}} \cdot \frac{d\bar{\zeta} \wedge d\zeta}{\zeta - z}.$$

On the other hand, since

$$\int_{\partial D(z,\varepsilon)} \frac{f(\zeta)d\zeta}{\zeta - z} = \int_{\partial D(z,\varepsilon)} \frac{f(\zeta) - f(z)}{\zeta - z} d\zeta + \int_{\partial D(z,\varepsilon)} \frac{f(z)}{\zeta - z} d\zeta,$$

and by the assumption that $f(\zeta) \in C^1(\bar{U})$, there exists a constant c, such that

$$|f(\zeta) - f(z)| < c|\zeta - z|$$

on $\partial D(z, \varepsilon)$. Therefore,

$$\left|\int_{\partial D(z,\varepsilon)} \frac{f(\zeta) - f(z)}{\zeta - z} d\zeta\right| < c\int_{\partial D(z,\varepsilon)} \left|\frac{\zeta - z}{\zeta - z}\right| |d\bar{\zeta}| = 2\pi\varepsilon c.$$

The above integral tends to zero as ε tends to zero. Since $\zeta = z + \varepsilon e^{i\theta}$ for $\zeta \in \partial D(z, \varepsilon)$, $0 \le \theta \le 2\pi$, we have that

$$\int_{\partial D(z,\varepsilon)} \frac{f(z)d\zeta}{\zeta - z} = f(z)\int_0^{2\pi} \frac{\varepsilon e^{i\theta} i d\theta}{\varepsilon e^{i\theta}} = 2\pi i f(z).$$

Thus

$$\int_{\partial U} \frac{f(\zeta)d\zeta}{\zeta - z} - 2\pi i f(z) = \iint_{U(z,\varepsilon} \frac{\partial f}{\partial \bar{\zeta}} \cdot \frac{d\bar{\zeta} \wedge d\zeta}{\zeta - z} + O(\varepsilon),$$

where $O(\varepsilon)$ satisfies the condition of $O(\varepsilon)/\varepsilon$ tends to a constant as ε tends to zero. We now let $\varepsilon \to 0$ in the above equation and (2.1) follows. This completes the prove.

The next theorem can be directly obtained from Theorem 2.1.

Theorem 2.2 ***(Cauchy Integral Formula)*** *Suppose $U \subseteq \mathbb{C}$ is a bounded region with C^1 boundary, $f(z)$ is holomorphic on U and $f(z) \in C^1(\bar{U})$. Then*

$$f(z) = \frac{1}{2\pi i} \int_{\partial U} \frac{f(\zeta)}{\zeta - z} d\zeta. \tag{2.2}$$

Following this theorem, we also have

Theorem 2.3 ***(Cauchy Integral Theorem)*** *Suppose $U \subseteq \mathbb{C}$ is a bounded region with C^1 boundary, $F(z)$ is holomorphic on U and $F(z) \in C^1(\bar{U})$. Then*

$$\int_{\partial U} F(\zeta)\, d\zeta = 0 \tag{2.3}$$

Proof Without loss of generosity, we assume that U contains the origin. Let $f(z) = zF(z)$ and substitute in (2.2). Let $z = 0$ in (2.2). Then we get (2.3).

In other words, Cauchy Integral Theorem can be derived from Cauchy Integral Formula. Of course, Theorem 2.3 can also be proved directly from Cauchy-Green's Theorem.

On the other hand, Cauchy Integral Formula can be derived from Cauchy Integral Theorem. Fix a point z_0 in U and consider $U_{z_0,\varepsilon}$ (defined as before) as the region. Let $F(z) = f(z)/(z - z_0)$. With a proof similar to Theorem 2.1, we can get Theorem 2.2. Thus, Theorem 2.2 and Theorem 2.3 are equivalent. And they are the corner stones of the theory of complex analysis.

The other important application of Cauchy-Green's Formula is in solving the one-dimensional $\bar{\partial}$-problem. This will be discussed in Chapter 3.

If ψ is a continuous function, then the *support* of ψ is defined to be the closure of the set on which ψ is not zero. It is denoted by $\operatorname{supp} \psi$.

Theorem 2.4 *(Solution of one dimensional $\bar{\partial}$-problem) Suppose $\psi(z) \in C^1(\mathbb{C})$ and $\operatorname{supp}\psi$ is compact. Let*

$$u(z) = \frac{-1}{2\pi i} \iint_{\mathbb{C}} \frac{\psi(\zeta)}{\zeta - z} \, d\bar{\zeta} \wedge d\zeta. \tag{2.4}$$

Then $u(z) \in C^1(\mathbb{C})$ and it is a solution of $\partial u(z)/\partial \bar{z} = \psi(z)$.

Proof Fix $z \in \mathbb{C}$ and let $\zeta - z = \xi$. Then

$$u(z) = \frac{-1}{2\pi i} \iint_{\mathbb{C}} \frac{\psi(\xi + z)}{\xi} \, d\bar{\xi} \wedge d\xi.$$

Since $1/\xi$ is integrable on any compact set, $u(z)$ is continuous. If $h \in \mathbb{R}$ and $h \neq 0$, then

$$\frac{u(z+h) - u(z)}{h} = \frac{-1}{2\pi i} \iint_{\mathbb{C}} \frac{1}{\xi} \cdot \frac{\psi(\xi + z + h) - \psi(\xi + z)}{h} \, d\bar{\xi} \wedge d\xi.$$

For fixed z and ξ,

$$\frac{\psi(\xi + z + h) - \psi(\xi + z)}{h} \longrightarrow \frac{\partial \psi(\xi + z)}{\partial \xi}$$

as $h \to 0$.

Since $\psi \in C^1(\mathbb{C})$ and $\operatorname{supp}\psi$ is compact,

$$\frac{\psi(\xi + z + h) - \psi(\xi + z)}{h}$$

converges to

$$\partial \psi(\xi + z)/\partial \xi$$

uniformly for ξ and z. Since $1/|\xi|$ is integrable on any compact set, we have

$$\begin{aligned}
\frac{\partial u}{\partial x}(z) &= \lim_{h \to 0} \frac{1}{h}(u(z+h) - u(z)) \qquad (2.5)\\
&= \frac{-1}{2\pi i} \iint_{\mathbb{C}} \frac{1}{\xi} \frac{\partial \psi}{\partial \alpha}(\xi + z) \, d\bar{\xi} \wedge d\xi \\
&= \frac{-1}{2\pi i} \iint_{\mathbb{C}} \frac{\partial \psi(\zeta)}{\partial \alpha} \frac{1}{\zeta - z} \, d\bar{\zeta} \wedge d\zeta
\end{aligned}$$

where $\zeta = \alpha + i\beta$, $\alpha, \beta \in \mathbb{R}$. The function $\partial u/\partial x$ is continuous since the above limit is uniform for z in any compact set of $\mathbb{C}$. Similarly,

$$\begin{aligned}\frac{\partial u}{\partial y}(z) &= \frac{-1}{2\pi i}\iint_{\mathbb{C}} \frac{1}{\xi}\frac{\partial \psi}{\partial \beta}(\xi + z)\, d\bar{\xi} \wedge d\xi \\ &= \frac{-1}{2\pi i}\iint_{\mathbb{C}} \frac{\partial \psi(\zeta)}{\partial \beta}\frac{1}{\zeta - z}\, d\bar{\zeta} \wedge d\zeta,\end{aligned} \tag{2.6}$$

and $\partial u/\partial y$ is continuous. Thus, $u \in C^1(\mathbb{C})$.

From (2.5) and (2.6) we have

$$\frac{\partial u}{\partial \bar{z}} = \frac{-1}{2\pi i}\iint_{\mathbb{C}} \frac{\partial \psi(\zeta)}{\partial \bar{\zeta}}\frac{1}{\zeta - z}\, d\bar{\zeta} \wedge d\zeta. \tag{2.7}$$

Since $\operatorname{supp}\psi$ is compact, there exists an $R > 0$, such that $\operatorname{supp}\psi \subset D(0, R)$ where $D(0, R) = \{z \mid |z| < R\}$. Thus, by (2.7) we have

$$\frac{\partial u}{\partial \bar{z}} = \frac{-1}{2\pi i}\iint_{D(0,R+\varepsilon)} \frac{\partial \psi}{\partial \bar{\zeta}}\frac{1}{\zeta - z}\, d\bar{\zeta} \wedge d\zeta$$

for $\varepsilon > 0$. By Cauchy-Green's Formula, the right hand side of the above equation is equal to

$$\psi(z) - \frac{1}{2\pi i}\int_{\partial D(0,R+\varepsilon)} \frac{\psi(\zeta)}{\zeta - z}\, d\zeta.$$

Obviously,

$$\frac{1}{2\pi i}\int_{\partial D(0,R+\varepsilon)} \frac{\psi(\zeta)}{\zeta - z}\, d\zeta = 0.$$

Therefore

$$\frac{\partial u(z)}{\partial \bar{z}} = \psi(z).$$

This proves the theorem.

It is easy to see that, if the support of $\psi(z) \in C^k(\mathbb{C})$ is compact, then $u(z)$ defined by (2.4) is also C^k. Similarly, if the support of $\psi(z) \in C^k(\mathbb{C})$ is a union of mutually disjoint compact sets (finitely many or infinitely many), then Theorem 2.4 still holds.

2.2 Cauchy-Goursat Theorem

Theorem 2.2 and Theorem 2.3 are the original form of the integral formula and integral theorem developed by Cauchy. Later on, the condition that $f(z) \in C^1(\bar{U})$ was dropped by Goursat and the two theorems were generalized.

Theorem 2.5 *(Cauchy-Goursat Integral Formula)* *Suppose $U \subset \mathbb{C}$ is a bounded region and ∂U is a simple closed curve. If $f(z)$ is holomorphic on U and continuous on $\bar{U}$, then*

$$f(z) = \frac{1}{2\pi i} \int_{\partial U} \frac{f(\zeta)}{\zeta - z} \, d\zeta. \tag{2.8}$$

Theorem 2.6 *(Cauchy-Goursat Integral Theorem)* *Suppose $U \subset \mathbb{C}$ is a bounded region and ∂U is a simple closed curve. If $f(z)$ is holomorphic on U and continuous on $\bar{U}$, then*

$$\int_{\partial U} f(\zeta) \, d\zeta = 0. \tag{2.9}$$

The two theorems above are equivalent. We only prove the later by a traditional argument.

Lemma 2.1 *Suppose $f(z)$ is a continuous function on a region $G \subseteq \mathbb{C}$, Γ is a piecewise smooth curve in G. Then for any $\varepsilon > 0$, there exists a broken line P inscribing Γ and contained in G, such that*

$$\left| \int_\Gamma f(z) \, dz - \int_P f(z) \, dz \right| < \varepsilon.$$

Proof Let $\bar{D} \subset G$ be the closure of a region such that $\Gamma \subset \bar{D}$. Since $f(z)$ is continuous on G, it is uniformly continuous on $\bar{D}$. Thus, for any $\varepsilon > 0$, there exists a $\delta = \delta(\varepsilon)$, such that $|f(z') - f(z'')| < \varepsilon$ for any z' and z'' in $\bar{D}$ that satisfy $|z' - z''| < \delta$. Divide Γ into n pieces of arcs $s_0, s_1, \cdots, s_{n-1}$ such that the length of each arc is less than δ. Draw a broken line P inscribing Γ, where P consists of the line segments $l_0, l_1, \cdots, l_{n-1}$, and z_i, z_{i+1} are the common end points of l_i and s_i for $i = 0, 1, \cdots, n-1$. Since the length of each s_k is less than δ, the distance of any two points on each arc is less than δ as well as the distance of any two points on l_k. The integral

$$\int_\Gamma f(z) \, dz$$

has an approximate value

$$s = f(z_0)\Delta z_0 + f(z_1)\Delta z_1 + \cdots + f(z_{n-1})\Delta z_{n-1},$$

where

$$\Delta z_k = \int_{s_k} dz.$$

So s can also be written as

$$s = \int_{s_0} f(z_0)\,dz + \int_{s_1} f(z_1)\,dz + \cdots + \int_{s_{n-1}} f(z_{n-1})\,dz.$$

Thus

$$\begin{aligned}\int_\Gamma f(z)dz - s = & \int_{s_0} (f(z) - f(z_0))\,dz + \int_{s_1} (f(z) - f(z_1))\,dz \\ & + \cdots + \int_{s_{n-1}} (f(z) - f(s_{n-1}))\,dz.\end{aligned}$$

Since $|f(z) - f(z_k)| < \varepsilon$ on every s_k, we have

$$\left| \int_\Gamma f(z)\,dz - s \right| < \varepsilon s_0 + \varepsilon s_1 + \cdots + \varepsilon s_{n-1} = \varepsilon l,$$

where l is the length of Γ.

Since Δz_k is also equal to

$$\int_{l_k} dz,$$

similarly, we have

$$\begin{aligned}\int_P f(z)\,dz - s = & \int_{l_0} (f(z) - f(z_0))\,dz + \int_{l_1} (f(z) - f(z_1))\,dz \\ & + \cdots + \int_{l_{n-1}} (f(z) - f(z_{n-1}))\,dz.\end{aligned}$$

Thus

$$\begin{aligned}\left| \int_P f(z)\,dz - s \right| & < \varepsilon l_0 + \varepsilon l_1 + \cdots + \varepsilon l_{n-1} \\ & = \varepsilon(l_0 + l_1 + \cdots + l_{n-1}) < \varepsilon l.\end{aligned}$$

Therefore

$$\left|\int_\Gamma f(z)\,dz - \int_P f(z)\,dz\right| \le \left|\int_\Gamma f(z)\,dz - s\right| + \left|s - \int_P f(z)\,dz\right|$$
$$= \varepsilon l + \varepsilon l = 2\varepsilon l.$$

This proves Lemma 2.1.

Lemma 2.2 *If $f(z)$ is a holomorphic function on a simply connected region $G \subseteq \mathbb{C}$, then*

$$\int_\Gamma f(\zeta)\,d\zeta = 0$$

where Γ is a piecewise smooth closed curve in G.

Proof By Lemma 2.1, any piecewise smooth closed curve Γ can be inscribed by a broken line P and

$$\left|\int_\Gamma f(z)\,dz - \int_P f(z)\,dz\right| < \varepsilon$$

for any $\varepsilon > 0$.

Lemma 2.2 will hold for any piecewise smooth curve if for any closed broken line P,

$$\int_P f(z)\,dz = 0.$$

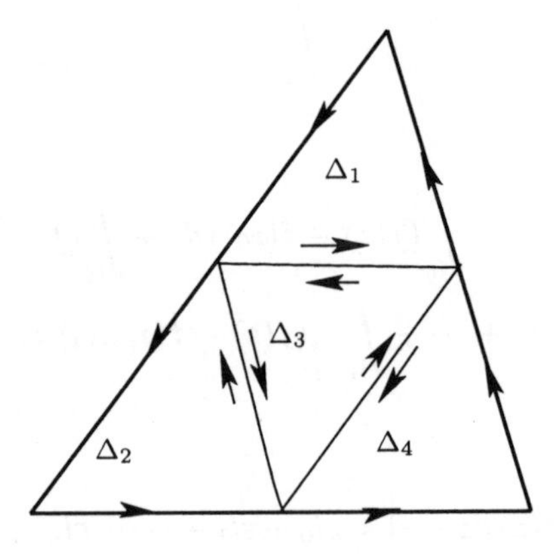

Fig. 1

We can always add some line segments to triangulate the interior of any closed broken lines. Since the value of integrals can cancel each other out on the added line segment, the integral on the closed broken line is equal to the sum of the integral on each triangle. Thus, if the assertion that Lemma 2.2 holds on the triangles (Figure 1) can be proved, then Lemma 2.2 holds for any piecewise smooth curves.

Now we show that Lemma 2.2 holds for triangles.

Let M be the absolute value of the integral of $f(z)$ on the boundary of a triangle in G. That is

$$\left|\int_{\Delta} f(z)\,dz\right| = M.$$

We need to show that $M = 0$. We bisect each side of the triangle and connect the midpoints of each side. The given triangle is divided into four identical smaller triangles and their boundaries are denoted by Δ_1, Δ_2, Δ_3 and Δ_4 respectively. Thus

$$\int_{\Delta} f(z)\,dz = \left(\int_{\Delta_1} + \int_{\Delta_2} + \int_{\Delta_3} + \int_{\Delta_4}\right) f(z)\,dz.$$

Since

$$\left|\int_{\Delta} f(z)\,dz\right| = M,$$

at least

$$\left|\int_{\Delta_k} f(z)\,dz\right| \geq \frac{M}{4}$$

for one of the $\Delta_k, (k = 1, 2, 3, 4)$. Let Δ_1 be such a triangle and we call it $\Delta^{(1)}$. Then

$$\left|\int_{\Delta^{(1)}} f(z)\,dz\right| \geq \frac{M}{4}.$$

Divide Δ_1 into four identical smaller triangles by the same method. We can find $\Delta^{(2)}$ such that

$$\left|\int_{\Delta^{(2)}} f(z)\,dz\right| \geq \frac{M}{4^2}.$$

Continuing this process, we can find a sequence of triangle $\Delta = \Delta^{(0)}$, $\Delta^{(1)}$, $\Delta^{(2)}, \cdots, \Delta^{(n)}, \cdots$, such that each one contains the succeeding one, and

$$\left|\int_{\Delta^{(n)}} f(z)\,dz\right| \geq \frac{M}{4^n}, \quad n = 0, 1, 2, \cdots. \tag{2.10}$$

Let L be the circumference of Δ. Then the circumference of $\Delta^{(1)}$, $\Delta^{(2)}$, $\cdots$, $\Delta^{(n)}$, $\cdots$ are $L/2$, $L/2^2$, $\cdots$, $L/2^n$, $\cdots$ respectively. This sequence tends to zero as n tends to infinity. Hence, a point z_0 can be found in all $\Delta^{(n)}$ $(n = 0, 1, 2, \cdots)$ such that for any $\varepsilon > 0$, there exists a $\delta = \delta(\varepsilon)$, such that

$$\left|\frac{f(z) - f(z_0)}{z - z_0} - f'(z_0)\right| < \varepsilon$$

for $|z - z_0| < \delta$, or equivalently

$$|f(z) - f(z_0) - f'(z_0)(z - z_0)| < \varepsilon|z - z_0|.$$

For a sufficiently large positive integer N, every $\Delta^{(n)}$ is contained in $D(z_0, \delta)$ for $n \geq N$. Obviously,

$$\int_{\Delta^{(n)}} dz = 0$$

and

$$\int_{\Delta^{(n)}} z\,dz = 0.$$

Thus

$$\int_{\Delta^{(n)}} f(z)\,dz = \int_{\Delta^{(n)}} (f(z) - f(z_0) - (z - z_0)f'(z_0))dz.$$

Therefore

$$\left|\int_{\Delta^{(n)}} f(z)\,dz\right| < \int_{\Delta^{(n)}} \varepsilon|z - z_0||dz|.$$

Since $|z - z_0|$ is the distance from a point z on $\Delta^{(n)}$ to a point z_0 in the interior of $\Delta^{(n)}$, we have

$$|z - z_0| < \frac{L}{2^n}.$$

Hence

$$\left|\int_{\Delta^{(n)}} f(z)\,dz\right| < \varepsilon\frac{L}{2^n} \cdot \frac{L}{2^n} = \varepsilon\frac{L^2}{4^n} \tag{2.11}$$

Comparing (2.10) with (2.11), we see that $M < \varepsilon L^2$ for any $\varepsilon > 0$. It follows that $M = 0$ and Lemma 2.2 is proved.

Proof of Theorem 2.6 First, we assume that U has a special shape. Let ∂U consists of $x = a$, $x = b\,(a < b)$ and two rectifiable continuous curves (Figure 2):

$$MN: y = \varphi(x) \quad (a \le x \le b), \quad PQ: y = \phi(x) \quad (a \le x \le b)$$

where $\varphi(x) < \phi(x)\,(a < x < b)$. Suppose $f(z)$ is holomorphic on U and continuous on $\bar{U}$. We want to show that

$$\int_{MNQPM} f(z)\,dz = 0. \tag{2.12}$$

We draw two straight lines $x = a + \varepsilon$, $x = b - \varepsilon$, and curves

$$M'N': y = \varphi(x) + \eta, \quad a \le x \le b; \quad P'Q': y = \phi(x) - \eta, \quad a \le x \le b,$$

where ε, η are sufficiently small positive numbers. Since U is a simply connected region, we have

$$\int_{M_1'N_1'Q_1'P_1'M_1'} f(z)\,dz = 0,$$

where $M_1'N_1'Q_1'P_1'M_1'$ is the boundary of the region which bounded by the two straight lines and the two curves defined above.

Fix ε and let $\eta \to 0$. Since $f(z)$ is uniformly continuous on $\bar{U}$, we have

$$\int_{M_1'N_1'} f(z)\,dz \to \int_{M_1N_1} f(z)\,dz, \quad \int_{Q_1'P_1'} f(z)\,dz \to \int_{Q_1P_1} f(z)\,dz,$$

$$\int_{P_1'M_1'} f(z)\,dz \to \int_{P_1M_1} f(z)\,dz, \quad \int_{N_1'Q_1'} f(z)\,dz \to \int_{N_1Q_1} f(z)\,dz.$$

Thus

$$\int_{M_1N_1Q_1P_1M_1} f(z)\,dz = 0.$$

Similarly, let $\varepsilon \to 0$, we have

$$\int_{M_1N_1} f(z)\,dz \to \int_{MN} f(z)\,dz, \quad \int_{Q_1P_1} f(z)\,dz \to \int_{QP} f(z)\,dz.$$

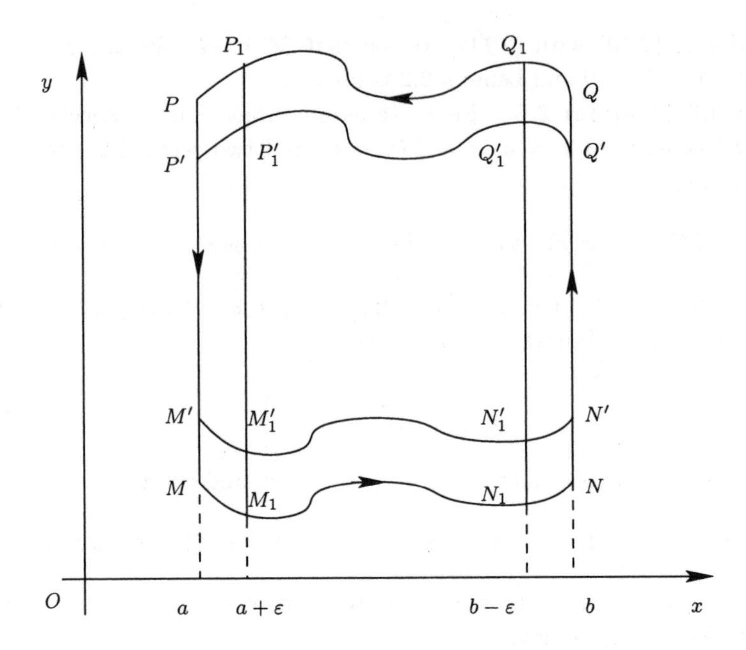

Fig. 2

If we can show that

$$\int_{P_1M_1} f(z)\,dz \to \int_{PM} f(z)\,dz, \quad \int_{N_1Q_1} f(z)\,dz \to \int_{NP} f(z)\,dz$$

as $\varepsilon \to 0$, then (2.12) can be proved. We only prove the second limit here since the proof of the first limit can be processed in a similar way.

Let

$$y_\varepsilon = \max(\varphi(b), \varphi(b-\varepsilon)), \quad Y_\varepsilon = \max(\phi(b), \phi(b-\varepsilon)).$$

Then

$$\begin{aligned}\int_{NQ} f(z)\,dz &= i\int_{\varphi(b)}^{\phi(b)} f(b+iy)\,dy \\ &= i\left(\int_{\varphi(b)}^{y_\varepsilon} + \int_{y_\varepsilon}^{Y_\varepsilon} + \int_{Y_\varepsilon}^{\phi(b)}\right) f(b+iy)\,dy,\end{aligned}$$

$$\int_{N_1Q_1} f(z)\,dz + i\int_{\varphi(b-\varepsilon)}^{\phi(b-\varepsilon)} f(b-\varepsilon+iy)\,dy$$
$$= i\left(\int_{\varphi(b-\varepsilon)}^{y_\varepsilon} + \int_{y_\varepsilon}^{Y_\varepsilon} + \int_{Y_\varepsilon}^{\phi(b-\varepsilon)}\right) f(b-\varepsilon+iy)\,dy.$$

Thus

$$\int_{NQ} f(z)\,dz - \int_{N_1Q_1} f(z)\,dz$$
$$= i\int_{y_\varepsilon}^{Y_\varepsilon} (f(b+iy) - f(b-\varepsilon+iy))\,dy + iS(\varepsilon), \qquad (2.13)$$

where

$$S(\varepsilon) = \left(\int_{\varphi(b)}^{y_\varepsilon} + \int_{Y_\varepsilon}^{\phi(b)}\right) f(b+iy)\,dy - \left(\int_{\varphi(b-\varepsilon)}^{y_\varepsilon} + \int_{Y_\varepsilon}^{\phi(b-\varepsilon)}\right) f(b-\varepsilon+iy)\,dy.$$

Since $f(z)$ is uniformly continuous, the first term on the right hand side of (2.13) tends to zero as ε tends to zero. Moreover, $\varphi(b)$, $\phi(b)$ are the limits of y_ε, Y_ε as $\varepsilon \to 0$. It follows that the four integrals in $S(\varepsilon)$ are all equal to zero. Hence

$$\int_{N_1Q_1} f(z)\,dz \to \int_{NQ} f(z)\,dz$$

as $\varepsilon \to 0$.

This shows that Theorem 2.6 holds when U is the special region above. Since any region can be divided into smaller regions of the above shape by finitely many auxiliary lines parallel to the y-axis, and the integrals on the auxiliary lines cancel each other, the theorem follows.

Theorem 2.6 is also true for multiply connected regions since a multiply connected region can be viewed as the union of some simply connected regions bounded by several auxiliary lines and the integrals on the auxiliary lines cancel each other. In other words, let $\gamma_0, \gamma_1, \cdots, \gamma_n$ be $n+1$ rectifiable closed curves. Let the curve γ_0 contain $\gamma_1, \gamma_2, \cdots, \gamma_n$ but γ_i lie on the outside of γ_j for $i \neq j$, $i, j = 1, 2, \cdots, n$. Let U be a region bounded by $\gamma_0, \gamma_1, \cdots, \gamma_n$, that is, ∂U consists of $\gamma_0, \gamma_1, \gamma_2, \cdots, \gamma_n$. If $f(z)$ is holomorphic on U and continuous on $\bar{U}$, then

$$\int_{\partial U} f(z)\,dz = 0.$$

By Theorem 2.6, if $f(z)$ is holomorphic on a simply connected region U and z_0, z are two points in U, we can define the integral of $f(z)$ as

$$F(z) = \int_{z_0}^{z} f(\zeta)\, d\zeta.$$

This integral is independent of the choice of the path. Obviously $F'(z) = f(z)$.

2.3 Taylor Series and Liouville Theorem

In this and the following sections, we will talk about several important corollaries of Cauchy Integral Formula and Cauchy Integral Theorem.

Theorem 2.7 *If $f(z)$ is holomorphic on a region $U \subseteq \mathbb{C}$ and continuous on $\bar{U}$, then for each point in U, the n-th derivative of $f(z)$ exists for all $n \in \mathbb{N}$ and*

$$f^{(n)}(z) = \frac{n!}{2\pi i} \int_{\partial U} \frac{f(\zeta)}{(\zeta - z)^{n+1}}\, d\zeta, \quad n = 1, 2, \cdots. \tag{2.14}$$

If $z_0 \in U$ and $\bar{D}(z_0, r) = \overline{\{z \,|\, |z - z_0| < r\}} \subset U$, then $f(z)$ has the Taylor expansion

$$f(z) = \sum_{j=0}^{\infty} a_j (z - z_0)^j \tag{2.15}$$

on $D(z_0, r)$. This series converges absolutely and uniformly in $\bar{D}(z_0, r)$, and

$$a_j = \frac{1}{2\pi i} \int_{\partial U} \frac{f(\zeta) d\zeta}{(\zeta - z_0)^{j+1}}. \tag{2.16}$$

Proof Let $z_0 \in U$ and the disc $D(z_0, r) \subset U$. By Theorem 2.5, if $z \in D(z_0, r)$, then

$$f(z_0) = \frac{1}{2\pi i} \int_{\partial U} \frac{f(\zeta)}{\zeta - z_0}\, d\zeta,$$

$$f(z) = \frac{1}{2\pi i} \int_{\partial U} \frac{f(\zeta)}{\zeta - z}\, d\zeta.$$

Thus

$$\begin{aligned} f(z) - f(z_0) &= \frac{1}{2\pi i} \int_{\partial U} \left(\frac{1}{\zeta - z} - \frac{1}{\zeta - z_0} \right) f(\zeta)\, d\zeta \\ &= \frac{z - z_0}{2\pi i} \int_{\partial U} \frac{f(\zeta) d\zeta}{(\zeta - z)(\zeta - z_0)}. \end{aligned}$$

It follows that

$$\frac{f(z) - f(z_0)}{z - z_0} = \frac{1}{2\pi i} \int_{\partial U} \frac{f(\zeta) d\zeta}{(\zeta - z)(\zeta - z_0)}.$$

Therefore

$$\begin{aligned} &\frac{f(z) - f(z_0)}{z - z_0} - \frac{1}{2\pi i} \int_{\partial U} \frac{f(\zeta) d\zeta}{(\zeta - z_0)^2} \\ &= \frac{1}{2\pi i} \int_{\partial U} \frac{f(\zeta)}{\zeta - z_0} \left(\frac{1}{(\zeta - z)} - \frac{1}{\zeta - z_0} \right) d\zeta \\ &= \frac{z - z_0}{2\pi i} \int_{\partial U} \frac{f(\zeta) d\zeta}{(\zeta - z)(\zeta - z_0)^2}. \end{aligned} \tag{2.17}$$

Let d be the distance from z_0 to ∂U and $r = d/2$. Then

$$|\zeta - z| = |(\zeta - z_0) - (z - z_0)| \geq |\zeta - z_0| - |z - z_0| \geq d - \frac{d}{2} = \frac{d}{2}.$$

Since $|\zeta - z_0| \geq d$, we have

$$\left| \int_{\partial U} \frac{f(\zeta) d\zeta}{(\zeta - z)(\zeta - z_0)^2} \right| \leq \frac{ML}{(d/2)d^2} = \frac{2ML}{d^3}$$

where $M = \max_{\zeta \in \partial U} |f(\zeta)|$ and L is the length of ∂U. Let $z \to z_0$ in (2.17). Then by the estimation above, we have

$$f'(z_0) = \frac{1}{2\pi i} \int_{\partial U} \frac{f(\zeta) d\zeta}{(\zeta - z_0)^2}.$$

This shows that (2.14) is true for $n = 1$.

Suppose

$$f^{(k)}(z) = \frac{k!}{2\pi i} \int_{\partial U} \frac{f(\zeta)}{(\zeta - z)^{k+1}}\, d\zeta$$

for $n = k \geq 1$.

Since

$$\frac{1}{\zeta - z} = \frac{1}{\zeta - z_0} \cdot \frac{1}{1 - \dfrac{z - z_0}{\zeta - z_0}},$$

and $|z - z_0| < r \le |\zeta - z_0|$, we have

$$\left| \frac{z - z_0}{\zeta - z_0} \right| < 1.$$

Thus

$$\frac{1}{\zeta - z} = \frac{1}{\zeta - z_0} \sum_{j=0}^{\infty} \left(\frac{z - z_0}{\zeta - z_0} \right)^j. \tag{2.18}$$

Therefore

$$\begin{aligned} f^{(k)}(z) &= \frac{k!}{2\pi i} \int_{\partial U} \frac{f(\zeta)}{(\zeta - z_0)^{k+1}} \left(1 + \frac{z - z_0}{\zeta - z_0} + \cdots \right)^{k+1} d\zeta \\ &= f^{(k)}(z_0) + \frac{(k+1)!}{2\pi i} \int_{\partial U} \frac{f(\zeta)(z - z_0) d\zeta}{(\zeta - z_0)^{k+2}} + O(|z - z_0|^2). \end{aligned}$$

It follows that

$$\frac{f^{(k)}(z) - f^{(k)}(z_0)}{z - z_0} = \frac{(k+1)!}{2\pi i} \int_{\partial U} \frac{f(\zeta) d\zeta}{(\zeta - z_0)^{k+2}} + O(|z - z_0|).$$

Let $z \to z_0$. We get

$$f^{(k+1)}(z_0) = \frac{(k+1)!}{2\pi i} \int_{\partial U} \frac{f(\zeta) d\zeta}{(\zeta - z_0)^{k+2}}.$$

By induction, (2.14) is ture for all $n \in \mathbb{N}$.

Substitute (2.18) into

$$f(z) = \frac{1}{2\pi i} \int_{\partial U} \frac{f(\zeta) d\zeta}{\zeta - z},$$

and note that by Theorem 1.15 in Chapter 1, the operations of integration and summation can be exchanged. Therefore, by (2.14), we have

$$\begin{aligned} f(z) &= \sum_{j=0}^{\infty} (z - z_0)^j \frac{1}{2\pi i} \int_{\partial U} \frac{f(\zeta) d\zeta}{(\zeta - z)^{j+1}} \\ &= \sum_{j=0}^{\infty} \frac{f^j(z_0)}{j!} (z - z_0)^j. \end{aligned}$$

This proves (2.15) and (2.16).

Theorem 2.7 demonstrates that, if the first order derivative of a complex function exists, then the n-th order derivative is also exists for all $n \in \mathbb{N}$. Moreover, the function can be expanded into the Taylor series. This property is not true for real functions and it indicates one of the essential differences between complex and real functions. In Section 1.3, we defined that a complex function $f(z)$ is holomorphic on $U \subseteq \mathbb{C}$ if its derivative exits at every point of U. By Theorem 2.7, we can also define that $f(z)$ is holomorphic on U if $f(z)$ can be expanded to a convergent power series in a neighborhood of z. Obviously, the two definitions are equivalent.

From Theorem 2.7, we have

Theorem 2.8 *(1) **(Cauchy Inequality)** If $f(z)$ is holomorphic on $U \subseteq \mathbb{C}$, and $\bar{D}(z_0, R) \subseteq U$, then*

$$|f^{(j)}(z_0)| \leq \frac{j!M}{R^j} \quad (j = 1, 2, \cdots) \tag{2.19}$$

where

$$M = \max_{z \in \bar{D}(z_0, R)} |f(z)|.$$

(2) Let $U \subseteq \mathbb{C}$ be a region, K be a compact subset of U and V be a neighborhood of K which is relatively compact in U (i.e $\bar{V}$ is a compact subset of U). Then for every holomorphic function $f(z)$ in U, there exist constants c_n $(n = 1, 2, \cdots)$ such that

$$\sup_{z \in K} |f^{(n)}(z)| \leq c_n \|f\|_{L(V)} \quad (n = 1, 2, \cdots), \tag{2.20}$$

where $\|f\|_{L(V)}$ is the L-norm of f on V, that is,

$$\frac{1}{A(V)} \iint_V |f(\zeta)| \, dA$$

and $A(V)$ is the area of V.

Cauchy inequality gives estimations for the modulus of derivatives of any order of holomorphic functions at a point. The inequality (2.20) gives estimations for the modulus of derivatives of any order of holomorphic functions on a compact set.

Proof of Theorem 2.8 Part (1) is obviously true. We now prove part (2).

Let ψ be a C^∞ function on V such that $\operatorname{supp}\psi$ is compact and $\psi = 1$ on a neighborhood of K contained in V. (The existence of ψ can be found in the appendix of this chapter.) Applying Theorem 2.1 (Pompeiu Formula) on ψf, we have

$$\psi(z)f(z) = \frac{1}{2\pi i}\int_{\partial U}\frac{\psi(\zeta)f(\zeta)d\zeta}{\zeta - z} + \frac{1}{2\pi i}\iint_U \frac{\partial(\psi f)}{\partial\bar{\zeta}}\cdot\frac{d\zeta\wedge d\bar{\zeta}}{\zeta - z}.$$

Since f is holomorphic on U, we have

$$\frac{\partial(\psi f)}{\partial\bar{\zeta}} = f\frac{\partial\psi}{\partial\bar{\zeta}}.$$

Since $\operatorname{supp}\psi(\zeta)$ is contained in V and V is relatively compact in U, we have

$$\psi(z)f(z) = \frac{1}{2\pi i}\iint_U f\frac{\partial\psi}{\partial\bar{\zeta}}\cdot\frac{d\zeta\wedge d\bar{\zeta}}{\zeta - z}.$$

If the support of $\partial\psi/\partial\bar{\zeta}$ is K_1, then K_1 is a compact subset of V and the distance between K and K_1, $d(K, K_1)$, is greater than zero.

If $z \in K$, then

$$f(z) = \frac{1}{2\pi i}\iint_{K_1} f(\zeta)\frac{\partial\psi(\zeta)}{\partial\bar{\zeta}}\cdot\frac{d\zeta\wedge d\bar{\zeta}}{\zeta - z}.$$

The the n-th derivative of the above equation with respect to z is

$$f^{(n)}(z) = \frac{n!}{2\pi i}\iint_{K_1} f(\zeta)\frac{\partial\psi(\zeta)}{\partial\bar{\zeta}}\cdot\frac{d\zeta\wedge d\bar{\zeta}}{(\zeta - z)^{n+1}}.$$

Thus

$$|f^{(n)}(z)| \le \frac{n!}{2\pi}\iint_{K_1} |f(\zeta)|\left|\frac{\partial\psi(\zeta)}{\partial\bar{\zeta}}\right|\cdot\frac{d\zeta\wedge d\bar{\zeta}}{|\zeta - z|^{n+1}}.$$

Since $d(K, K_1) > 0$, there exists a c_1, such that

$$\frac{1}{|\zeta - z|} < c_1$$

for any $z \in K$ and $\zeta \in K_1$. Obviously,

$$\left|\frac{\partial\psi(\zeta)}{\partial\bar{\zeta}}\right|$$

is bounded on K_1, so there exists a c'_n such that

$$|f^{(n)}(z)| \leq c'_n \iint_{K_1} |f(\zeta)|\,|d\zeta \wedge d\bar{\zeta}|$$
$$\leq c'_n \iint_V |f(\zeta)|\,|d\zeta \wedge d\bar{\zeta}| = c_n \|f\|_{L(V)}$$

where c'_n and c_n are constants which depend only on n. This proves (2.19).

Following by Theorem 2.8(2), we have

Corollary 2.1 *Suppose $U \subseteq \mathbb{C}$ is a region, K is a compact subset of U, $V \subset U$ is a neighborhood of K and is relatively compact in U. Then for any holomorphic function $f(z)$ in U, there exist constants c_n $(n = 1, 2, \cdots)$ which are independent of z such that*

$$\sup_{z \in K} |f^n(z)| \leq c_n \sup_{z \in V} |f(z)|, \quad n = 1, 2, \cdots.$$

This corollary can also be derived directly from Theorem 2.7, but Theorem 2.8(2) cannot be proved by Theorem 2.7. Theorem 2.8(2) is more profound. From Theorem 2.7, we can also get the converse of Theorem 2.6 (Cauchy-Goursat Integral Theorem).

Theorem 2.9 *(Morera Theorem)* *If $f(z)$ is continuous on U and the integral of f along any rectifiable closed curve is zero, then $f(z)$ is holomorphic on U.*

Proof We choose a point $z_0 \in U$, since the integral of f along any rectifiable closed curve is zero, the integral

$$F(z) = \int_{z_0}^{z} f(\zeta)\, d\zeta, \quad z \in U$$

is independent of the choice of path and $F'(z) = f(z)$. Hence $F(z)$ is holomorphic on U. By Theorem 2.7, the second order derivative of $F(z)$, $f'(z)$, also exists. Therefore, $f(z)$ is holomorphic on U and the theorem follows.

We can also get the following important theorems from Theorem 2.7.

Theorem 2.10 *(Liouville Theorem)* *If $f(z)$ is holomorphic and bounded on the entire complex plane $\mathbb{C}$, then f is a constant function.*

Proof Suppose $|f(z)| \leq M$ for $z \in \mathbb{C}$. Let $z_0 \in \mathbb{C}$ be a fixed point and $D(z_0, R)$ be a disc. By the case of $j = 1$ in (2.19), we have $|f'(z_0)| \leq M/R$. It follows that $f'(z_0) = 0$ if we let $R \to \infty$. Since z_0 is arbitrary in $\mathbb{C}$, $f'(z) = 0$ for any $z \in \mathbb{C}$. Hence $f(z)$ is a constant function on $\mathbb{C}$.

Liouville theorem demonstrates that the only functions that are holomorphic and bounded on the entire complex plane are the constant functions. Further discussion about this theorem can be found in Chapter 5.

Theorem 2.11 ***(Riemann Theorem)*** *Suppose $\tilde{D}(z_0, r) = D(z, r) \setminus \{z_0\}$, F is holomorphic and bounded on $\tilde{D}(z_0, r)$. Then F can be analytically continued to $D(z_0, r)$. In other words, there exists a holomorphic function f defined on $D(z_0, r)$ such that*

$$f|_{\tilde{D}(z_0,r)} = F$$

Proof Without loss of generosity, assume that $z_0 = 0$. Define

$$G(z) = \begin{cases} z^2 F(z), & z \in \tilde{D}(0, r), \\ 0, & z = 0. \end{cases}$$

Then $G(z)$ is continuously differentiable on $D(0, r)$ and satisfies Cauchy-Riemann equation since

$$\lim_{z \to 0} \frac{G(z) - 0}{z} = \lim_{z \to 0} \frac{z^2 F(z) - 0}{z} = \lim_{z \to 0} z F(z) = 0,$$

therefore $dG(0)/dz = 0$. If $z \neq 0$, then

$$G'(z) = z^2 F'(z) + 2z F(z).$$

It is easy to see that $G'(z) \to 0$ as $z \to 0$. By Theorem 1.12 of Section 1.3, $G(z)$ is holomorphic on $D(0, r)$. Hence $G(z)$ can be expanded to the Taylor series

$$G(z) = 0 + 0 \cdot z + a_2 z^2 + a_3 z^3 + \cdots \tag{2.21}$$

and it is convergent in $D(0, r)$. Define

$$f(z) = \frac{G(z)}{z^2} = a_2 + a_3 z + \cdots . \tag{2.22}$$

By (1.20) in Theorem 1.14 of Chapter 1, series (2.22) and series (2.21) have the same radius of convergence. Therefor, $f(z)$ is holomorphic on $D(0, r)$ and $f(z) = F(z)$ in $\tilde{D}(0, r)$.

2.4 Some Results about the Zeros of Holomorphic Functions

Let $f(z)$ be a holomorphic function on a region $U \subseteq \mathbb{C}$. A point z_0 is called a *zero* of $f(z)$ if $f(z) = 0$. If $f(z)$ has a power series expansion at $z = z_0$

$$a_m(z - z_0)^m + a_{m+1}(z - z_0)^{m+1} + \cdots, \quad a_m \neq 0$$

then we say that $f(z)$ has a zero at $z = z_0$ with multiplicity m. A series of corollaries about zeros can be obtained from Cauchy integral formula and Cauchy integral theorem.

Theorem 2.12 *(The Fundamental Theorem of Algebra)* *Let $p(z) = a_0 + a_1 z + \cdots + a_n z^n$ be an n-th degree polynomial. Then there exists at least one z_0, such that $p(z_0) = 0$. The number z_0 is called a root of the equation $p(z) = 0$.*

Proof Suppose the contrary. Then $f(z) = 1/p(z)$ is holomorphic on $\mathbb{C}$. Since $p(z) \to \infty$ as $z \to \infty$, $f(z)$ is bounded on $\mathbb{C}$. By Theorem 2.10 (Liouville Theorem), $f(z)$ is a constant function, and so is $p(z)$. This is a contradiction.

Theorem 2.13 *If $f(z)$ is holomorphic on a region $U \subseteq \mathbb{C}$, then the set of zeros of $f(z)$, $\{z \in U | f(z) = 0\}$, has no accumulation points in U unless $f(z)$ equal to zero identically.*

Proof Suppose the contrary. Let $z_1, z_2, \cdots, z_n, \cdots$ be the zeros of $f(z)$ on U. Assume this sequence of zeros has an accumulation point $z_0 \in U$. Without loss of generosity, let $z_0 = 0$. Since $f(z)$ is holomorphic on U and $0 \in U$, $f(z)$ can be expanded to the Taylor series

$$f(z) = a_0 + a_1 z + a_2 z^2 + \cdots$$

at point 0. Since every z_n in the sequence $\{z_n\}$ is a zero of $f(z)$, we have $f(z_n) = 0$ for all n. Thus

$$\lim_{n\to\infty} f(z_n) = f(\lim_{n\to\infty} z_n) = f(0) = 0.$$

Therefore, $a_0 = 0$ and

$$f(z) = a_1 z + a_2 z^2 + a_3 z^3 + \cdots.$$

It follows that $a_1 = f(z)/z + O(z)$. Let $z = z_n$. Then $a_1 = f(z_n)/z_n + O(z_n) = O(z_n)$. Hence $a_1 = 0$ when we let $n \to \infty$. Similarly, we can prove $a_2 = a_3 = \cdots = a_n = \cdots = 0$. Since all of the coefficients of the

Taylor series are zero, $f(z) = 0$ follows. Therefore, If $f(z)$ is not equal to zero identically on U, then the set $\{z \in U | \, f(z) = 0\}$ has no accumulation points in U.

By Theorem 2.13, if $h_1(z)$ and $h_2(z)$ are holomorphic functions on a region $U \subseteq \mathbb{C}$, E is a subset of U which has an accumulation point of U and $h_2(z) = h_2(z)$ on E, then $h_1(z) = h_2(z)$ on U. In other words, the value of a holomorphic function on U can be determined by the value of the function on a subset of U that contains an accumulation point in U. For instance, $\sin^2 z + \cos^2 z = 1$ is ture for real z, it is also true for complex z. The other real trigonometric identities hold in the complex number field for the same reason.

Theorem 2.14 *(The Argument Principle)* *Let $f(z)$ be a holomorphic function on a region $U \subseteq \mathbb{C}$, $\gamma \subset U$ be a simple closed curve with positive orientation and homotopic to zero in U. Assume the function $f(z)$ does not vanish on γ. Then $f(z)$ has finitely many zeros in the interior of γ and the number of zeros k (counting multiplicity) is*

$$k = \frac{1}{2\pi i} \int_\gamma \frac{f'(z)}{f(z)}\, dz. \tag{2.23}$$

Let $w = f(z)$. Then

$$k = \frac{1}{2\pi i} \int_\gamma \frac{f'(z)}{f(z)}\, dz = \frac{1}{2\pi i} \int_\Gamma \frac{dw}{w} = \frac{1}{2\pi} \Delta_\Gamma \operatorname{Arg} w,$$

where Γ is the image of γ under the map f, and $\Delta_\Gamma \operatorname{Arg} w$ is the change of the argument of w on Γ.

This theorem tells us that when z travels along γ in the positive direction once, the number of circles that $w = f(z)$ travels around the origin along Γ in the positive direction is exactly the same as the number of zeros of f in the interior of γ. That is why this theorem is also called the argument principle.

Proof of Theorem 2.14 Suppose $f(z)$ has zeros $z_1, z_2, \cdots, z_n$ with multiplicity $k_1, k_2, \cdots, k_n$ respectively inside γ. Draw circles γ_i with center z_i and radius $\varepsilon_i > 0$ for $i = 1, 2, \cdots, n$, such that every γ_i lies inside of γ and they are mutually exclusive. By Cauchy Integral Formula, we have

$$\frac{1}{2\pi i} \int_\gamma \frac{f'(z)}{f(z)}\, dz = \sum_{i=1}^{n} \frac{1}{2\pi i} \int_{\gamma_i} \frac{f'(z)}{f(z)}\, dz.$$

Since the multiplicity of z_i is k_i, $f(z)$ can be written as

$$f(z) = (z - z_i)^{k_i} h_i(z),$$

for z inside γ_i, where $h_i(z) \neq 0$. Therefore

$$f'(z) = k_i(z - z_i)^{k_i - 1} h_i(z) + (z - z_i)^{k_i} h_i'(z).$$

It follows that

$$\frac{f'(z)}{f(z)} = \frac{k_i}{z - z_i} + \frac{h_i'(z)}{h_i(z)}.$$

Hence

$$\frac{1}{2\pi i} \int_{\gamma_i} \frac{f'(z)}{f(z)}\, dz = k_i,$$

and

$$\frac{1}{2\pi i} \int_{\gamma} \frac{f'(z)}{f(z)}\, dz = \sum_{i=1}^{n} k_i = k.$$

Theorem 2.15 ***(Hurwitz Theorem)*** *Let $\{f_j\}$ be a sequence of holomorphic functions on $U \subseteq \mathbb{C}$ that converges uniformly to a function f on every compact subset of U. If f_j is never equal to zero on U for all j, then f is either identically zero or never equal to zero on U.*

Proof For an arbitrary point $z \in U$, choose a simple closed curve γ in U such that the inside of γ contains z. Since f_j is holomorphic on U, by Cauchy integral formula we have

$$f_j(z) = \frac{1}{2\pi i} \int_{\gamma} \frac{f_j(\zeta) d\zeta}{\zeta - z}.$$

Since $\{f_j\}$ converges uniformly on every compact subset of U, we have

$$\lim_{j\to\infty} f_j(z) = \lim_{j\to\infty} \frac{1}{2\pi i} \int_{\gamma} \frac{f_j(\zeta) d\zeta}{\zeta - z} = \frac{1}{2\pi i} \int_{\gamma} \lim_{j\to\infty} f_j(\zeta) \frac{d\zeta}{\zeta - z}.$$

It follows that

$$f(z) = \frac{1}{2\pi i} \int_{\gamma} \frac{f(\zeta) d\zeta}{\zeta - z}.$$

Hence, $f(z)$ is a holomorphic function. Similarly, we can prove that $\{f_j'(z)\}$ converges uniformly to $f'(z)$ on every compact subset of U.

Remark The property of the Integral of the Limit for uniformly convergent series we used in the proof above can be directly obtained from Theorem 1.15. The proof is also similar to the proof of the corresponding theorem in calculus. We leave it to the reader to supply a proof.

If $f(z)$ is not identically zero, then by Theorem 2.13, the zeros of f are discrete. Let γ be a curve that does not pass through these zeros. Then

$$\frac{1}{2\pi i}\int_\gamma \frac{f_j'(\zeta)}{f_j(\zeta)}d\zeta \to \frac{1}{2\pi i}\int_\gamma \frac{f'(\zeta)}{f(\zeta)}\,d\zeta,$$

as $j \to \infty$.

By the assumption and Theorem 2.14, we have

$$\frac{1}{2\pi i}\int_\gamma \frac{f_j'(\zeta)}{f_j(\zeta)}\,d\zeta = 0.$$

Therefore

$$\frac{1}{2\pi i}\int_\gamma \frac{f'(\zeta)}{f(\zeta)}\,d\zeta = 0,$$

and $f(z)$ has no zero on U.

Theorem 2.16 ***(Rouché Theorem)*** *If $f(z)$ and $g(z)$ are holomorphic functions on $U \subseteq \mathbb{C}$, γ is a rectifiable simple closed curve that lies inside of U, and*

$$|f(z) - g(z)| < |f(z)| \tag{2.24}$$

for all $z \in \gamma$, then f and g have the same number of zeros enclosed by γ.

Proof By (2.24), we have $|f(z)| > 0$ and $g(z) \neq 0$ on γ. (If not, then there exists a point z_0 on γ, where $g(z_0) = 0$. It follows that $|f(z_0)| < |f(z_0)|$ and this is impossible.)

Let N_1 and N_2 be the number of zeros of f and g inside γ respectively. Then by Theorem 2.14 (The Argument Principle), we have

$$N_1 = \frac{1}{2\pi i}\int_\gamma \frac{f'(z)}{f(z)}\,dz, \quad N_2 = \frac{1}{2\pi i}\int_\gamma \frac{g'(z)}{g(z)}\,dz.$$

It follows that

$$\begin{aligned} N_2 - N_1 &= \frac{1}{2\pi i} \int_\gamma \left(\frac{g'(z)}{g(z)} - \frac{f'(z)}{f(z)} \right) dz \\ &= \frac{1}{2\pi i} \int_\gamma \frac{fg' - gf'}{fg} dz \\ &= \frac{1}{2\pi i} \int_\gamma \frac{(g/f)'}{g/f} dz. \end{aligned}$$

Let $F(z) = g(z)/f(z)$. Then

$$N_2 - N_1 = \frac{1}{2\pi i} \int_\gamma \frac{F'(z)}{F(z)} dz.$$

Note that (2.24) is equivalent to $|F(z) - 1| < 1$. The function $w = F(z)$ maps γ to Γ. Since $|w - 1| < 1$ on Γ, Γ does not pass the origin nor enclose the origin. By Theorem 2.3 (Cauchy Integral Theorem) we have

$$\int_\Gamma \frac{dw}{w} = 0$$

and $N_1 = N_2$ follows.

It was proved in Theorem 2.12 that the n-th degree polynomial $p(z) = a_n z^n + a_{n-1} z^{n-1} + \cdots + a_0$ has at least one zero.

Now, using Rouché Theorem, we can prove that if $a_n \neq 0$, then $p(z)$ has exactly n zeros.

Let $g(z) = a_n z^n$. If $|z| = R$ is sufficiently large, we have

$$\begin{aligned} |p(z) - g(z)| &= |a_{n-1} z^{n-1} + \cdots + a_0| < |g(z)| \\ &= |a_n||z|^n = |a_n||R|^n. \end{aligned}$$

By Rouché Theorem, $p(z)$ and $g(z)$ have the same number of zeros in $|z| < R$. Obviously, $a_n z^n$ has n zeros and so does $p(z)$.

As a corollary of Rouché Theorem, we have

Theorem 2.17 *Let $f(z)$ be a holomorphic function on a region $U \subseteq \mathbb{C}$, $w_0 = f(z_0)$ and $z_0 \in U$. If z_0 is a zero of $f(z) - w_0$ with multiplicity m, then for a sufficiently small $\rho > 0$, there exists a $\delta > 0$, such that for every point A in $D(w_0, \delta)$, the function $f(z) - A$ has exact m zeros in $D(z_0, \rho)$.*

Proof Since z_0 is a zero of $f(z) - f(z_0)$ with multiplicity m, by Theorem 2.13, there exists a $\rho > 0$, such that $f(z) - f(z_0)$ has no other zeros except

z_0 on $\bar{D}(z_0, \rho) \subseteq U$ and $|f(z) - f(z_0)| \geq \delta$ on $|z - z_0| = \rho$. Thus, for any point A in $D(w_0, \delta)$, $|A - w_0| < |f(z) - f(z_0)|$ on $|z - z_0| = \rho$. Hence

$$|(f(z) - f(z_0)) - (f(z) - A)| < |f(z) - f(z_0)|.$$

By Rouché Theorem, the number of zeros of $f(z) - A$ and $f(z) - f(z_0)$ are the same on $D(z_0, \rho)$. By the assumption, the multiplicity of zero of $f(z) - f(z_0)$ on $D(z_0, \rho)$ is m and it follows that $f(z) - A$ has m zeros on $D(z_0, \rho)$. This proves the theorem.

We will discuss more about the theory of zeros later.

2.5 Maximum Modulus Principle, Schwarz Lemma and Group of Holomorphic Automorphisms

Another important corollary of Cauchy Integral Formula is the Maximum Modulus Principle. It is a very useful theorem.

We need to prove the mean-value property of holomorphic functions first.

Let $f(z)$ be a holomorphic function on a region U, $z_0 \in U$, $r > 0$ and $\bar{D}(z_0, r) \subset U$. Then by Cauchy Integral Formula,

$$f(z_0) = \frac{1}{2\pi i} \int_{\partial D(z_0, r)} \frac{f(\zeta) d\zeta}{\zeta - z_0}.$$

Since the point ζ on $\partial D(z_0, r)$ can be written as $\zeta = z_0 + re^{it}$, where $0 \leq t \leq 2\pi$, the Cauchy Integral Formula becomes

$$\begin{aligned} f(z_0) &= \frac{1}{2\pi i} \int_0^{2\pi} \frac{f(z_0 + re^{it}) ire^{it}}{re^{it}} dt \\ &= \frac{1}{2\pi} \int_0^{2\pi} f(z_0 + re^{it})\, dt \end{aligned} \tag{2.25}$$

This is the mean-value property of holomorphic functions. The value of $f(z)$ at $z = z_0$ is equal to the mean of the values of $f(z)$ on $\partial D(z_0, r)$.

By taking the real and the imaginary parts of the both sides of (2.25), we find that harmonic functions also have the mean-value property. On the other hand, if a continuous real function satisfies the mean-value property, then it must be a harmonic function. This assertion will be discussed in Section 2.6. Since a real function has the mean-value property if and only if it is a harmonic function, harmonic functions can also be defined by the mean-value property.

Now, we prove the following theorem using the mean-value property of holomorphic functions.

Theorem 2.18 *(Maximum Modulus Principle)* *Let $f(z)$ be a holomorphic function on a region $U \subseteq \mathbb{C}$. If there exists a point $z_0 \in U$ such that $|f(z_0)| \geq |f(z)|$ for all $z \in U$, then $f(z)$ is a constant function.*

Proof We can multiply a number with modulus 1 on f so that $M = f(z_0) \geq 0$. Let $S = \{z \in U | f(z) = f(z_0)\}$. The set S is not empty since $z_0 \in S$ and is closed since f is continuous on U. Now we show that S is also open. Let $w \in S$, $r > 0$ and $D(w, r) \subseteq U$. By mean-value property, we have

$$
\begin{aligned}
M = f(w) &= \left| \frac{1}{2\pi} \int_0^{2\pi} f(w + r'e^{it})\, dt \right| \\
&\leq \frac{1}{2\pi} \int_0^{2\pi} |f(w + r'e^{it})|\, dt \leq M
\end{aligned}
$$

for $0 < r' < r$. Thus the equality holds and it follows that

$$
f(w + r'e^{it}) = |f(e + r'e^{it})| = M
$$

for all t and $0 < r' < r$. Therefore

$$
\{w + r'e^{it} | 0 \leq t \leq 2\pi, 0 < r' < r\} \subseteq S.
$$

In other words, for any point w in S, there exists an open disc such that every point of the disc is in S. So S is open. Hence, S is a non-empty, open and closed subset of U. Since U is simply connected, we have $S = U$. Therefore, $f(z)$ is a constant function on U and this proves the theorem.

The assertion that if a non-empty subset A of a connected set B is open and closed, then $A = B$ was used in the proof of the above theorem. It is a very useful result and we leave the verification to the reader.

As a corollary of the Maximum Modulus Principle, if $f(z)$ is a holomorphic function on a bounded region $U \subset \mathbb{C}$, continuous on $\bar{U}$ and not a constant function, then the maximum value of $|f(z)|$ can only be reached on ∂U.

In the proof of the Maximum Modulus Principle, only the mean-value property is used, so this principle is also true for harmonic functions.

The next important theorem can also be obtained from the Maximum Modulus Principle.

Theorem 2.19 ***(Schwarz Lemma)*** *If a holomorphic function $f(z)$ maps the unit disc $D = D(0,1)$ to itself and $f(0) = 0$, then*

$$|f(z)| \leq |z| \quad and \quad |f'(0)| \leq 1. \tag{2.26}$$

The equalities $|f(z)| = |z|$ and $|f'(0)| = 1$ hold for $z \neq 0$ if and only if $f(z) = e^{i\tau}z$, where $\tau \in \mathbb{R}$.

Proof Let

$$G(z) = \begin{cases} \dfrac{f(z)}{z}, & z \neq 0, \\ f'(0), & z = 0. \end{cases}$$

Then $G(z)$ is holomorphic on D. Apply Maximum Modulus Principle to $G(z)$ on $\{z \,|\, |z| \leq 1 - \varepsilon\}$ $(\varepsilon > 0)$ we get

$$|G(z)| \leq \frac{\max_{|z|=1-\varepsilon} |f(z)|}{1-\varepsilon} < \frac{1}{1-\varepsilon}.$$

Letting $\varepsilon \to 0^+$. We get $|G(z)| \leq 1$ on D. Thus, $|f(z)| \leq |z|$ for $z \neq 0$ and $|G(0)| = |f'(0)| \leq 1$.

If $|f(z)| = |z|$, or equivalently, $|G(z)| = 1$, at a point $z \neq 0$ in D, then by the Maximum Modulus Principle, $G(z)$ is a constant and $|G(z)| = 1$ for all $z \in D$. Thus $G(z) = e^{i\tau}$ and $f(z) = e^{i\tau}z$. Similarly, we can show that $f(z) = e^{i\tau}z$ if $|f'(0)| = 1$. The proof is complete.

From Schwarz Lemma, we can determine the group of holomorphic automorphisms of D.

Let U be a region contained in $\mathbb{C}$. The group of holomorphic automorphisms on U is defined as follows:

Suppose $f(z)$ is a holomorphic function defined on U. If $f(z)$ maps U one-to-one and holomorphically to itself, then $f(z)$ is called a *holomorphic automorphism* on U. All of the holomorphic automorphisms on U form a group, called the *group of holomorphic automorphisms* of U and is denoted by $\mathrm{Aut}(U)$.

Now we determine $\mathrm{Aut}(D)$.

First we show that if $a \in D$, then

$$\varphi_a(\zeta) = \frac{-\zeta + a}{1 - \bar{a}\zeta} \in \mathrm{Aut}(D).$$

It is easy to see that φ_a is holomorphic on $\bar{D}$ and $\varphi_a(a) = 0$. For $|\zeta| = 1$, we have

$$|\varphi_a(\zeta)| = \left|\frac{\zeta - a}{1 - \bar{a}\zeta}\right| = \left|\frac{1}{\bar{\zeta}} \cdot \frac{\zeta - a}{1 - \bar{a}}\zeta\right| = \left|\frac{\zeta - a}{\bar{\zeta} - \bar{a}}\right| = 1.$$

It follows that $\varphi_a(\zeta)$ maps the boundary of D to itself and thus it maps the inside of D to itself.

Next, we show that $\varphi_a(\zeta)$ is one-to-one on D.

Assume that $\zeta_1, \zeta_2 \in D$ and

$$\frac{-\zeta_1 + a}{1 - \bar{a}\zeta_1} = \frac{-\zeta_2 + a}{1 - \bar{a}\zeta_2}.$$

Then

$$(\zeta_1 - a)(1 - \bar{a}\zeta_2) = (\zeta_2 - a)(1 - \bar{a}\zeta_1).$$

It follows that $(\zeta_1 - \zeta_2)(1 - |a|^2) = 0$. This implies that $\zeta_1 = \zeta_2$ since $|a| < 1$. Therefore, $\varphi_a \in \mathrm{Aut}(D)$.

Let $\xi = \varphi_a(\zeta) = \dfrac{-\zeta + a}{1 - \bar{a}\zeta}$. Then

$$\xi - \bar{a}\zeta\xi = -\zeta + a, \quad \zeta = \frac{-\xi + a}{1 - \bar{a}\xi} = \varphi_a(\xi).$$

We have thus $(\varphi_a)^{-1} = \varphi_a$ and $\varphi_a \in \mathrm{Aut}(D)$. The map φ_a is referred to as a Möbius transformation. The group of all Möbius transformations is called the *Möbius transformation group* and it is a subgroup of $\mathrm{Aut}(D)$. Moreover, the rotation $\xi = \rho_\tau(\zeta) = e^{i\tau}\zeta, \tau \in \mathbb{R}$, is also in $\mathrm{Aut}(D)$. The group of all rotations is called the *rotation group* and it is also a subgroup of $\mathrm{Aut}(D)$.

Theorem 2.20 ***(The Group of Holomorphic Automorphisms on the Unit Disc)*** *If $f \in \mathrm{Aut}(D)$, then there exists $a \in \mathbb{C}$ with $|a| < 1$, and $\tau \in \mathbb{R}$, such that*

$$f(\zeta) = \varphi_a \circ \rho_\tau(\zeta). \tag{2.27}$$

In other words, every element of $\mathrm{Aut}(D)$ *is a composition of a Möbius transformation and a rotation.*

Proof Suppose $f(0) = b$ and $G = \varphi_b \circ f$. Then G is univalent and holomorphic on D. It maps D to D and

$$G(0) = \varphi_b \circ f(0) = \varphi_b(b) = 0$$

From Theorem 2.19 (Schwarz Lemma), we have $|G'(0)| \leq 1$. By the assumption of G, the inverse of G, G^{-1}, exists on D and is also univalent and holomorphic with $G^{-1}(0) = 0$. Similarly, by applying Theorem 2.19 on G^{-1}, we have

$$\left|\frac{1}{G'(0)}\right| = |(G^{-1})'(0)| \leq 1.$$

Thus $|G'(0)| = 1$. It follows that $G(\zeta) = e^{i\tau}\zeta = \rho_\tau(\zeta)$ and $\varphi_b \circ f = \rho_\tau$. Therefore, $f = \varphi_{-b} \circ \rho_\tau$. Let $-b = a$ and (2.27) follows.

The next important Lemma can be obtained from Theorem 2.20.

Theorem 2.21 *(Schwarz-Pick Lemma)* *Let f be a holomorphic function which maps D into D. If $z_1, z_2 \in D$, $w_1 = f(z_1)$ and $w_2 = f(z_2)$, then*

$$\left|\frac{w_1 - w_2}{1 - w_1\bar{w}_2}\right| \leq \left|\frac{z_1 - z_2}{1 - z_1\bar{z}_2}\right| \tag{2.28}$$

and

$$\frac{|dw|}{1 - |w|^2} \leq \frac{|dz|}{1 - |z|^2}, \tag{2.29}$$

where the equalities hold if and only if $f \in \mathrm{Aut}(D)$.

Proof Let

$$\varphi(z) = \frac{z + z_1}{1 + \bar{z}_1 z}, \quad \psi(z) = \frac{z - w_1}{1 - \bar{w}_1 z}.$$

It is easy to see that $\varphi, \psi \in \mathrm{Aut}(D)$, and

$$\psi \circ f \circ \varphi(0) = \psi \circ f(z_1) = \psi(w_1) = 0.$$

Thus $\psi \circ f \circ \varphi$ satisfies the condition of Theorem 2.19 (Schwarz Lemma). Therefore

$$|(\psi \circ f \circ \varphi)(z)| \leq |z|$$

for $z \in D$ and $z \neq 0$.

Let $z = \varphi^{-1}(z_2)$. Then

$$|\psi \circ f(z_2)| \leq |\varphi^{-1}(z_2)|.$$

It follows that $|\psi(w_2)| \leq |\varphi^{-1}(z_2)|$ and this is (2.28).

If $z = 0$, then by Theorem 2.19 we have

$$|(\psi \circ f \circ \varphi)'(0)| \leq 1,$$

that is

$$|\psi'(w_1)f'(z_1)\varphi'(0)| \leq 1.$$

Since

$$\varphi'(z) = \frac{1 - z_1\bar{z}_1}{(1 + \bar{z}_1 z)^2}, \quad \varphi'(0) = 1 - |z_1|^2,$$

$$\psi'(z) = \frac{1 - w_1\bar{w}_1}{(1 - \bar{w}_1 z)^2}, \quad \psi'(w_1) = \frac{1}{1 - |w_1|^2},$$

we have

$$|f'(z_1)| \leq \frac{1 - |w_1|^2}{1 - |z_1|^2},$$

and this is (2.29). By Theorem 2.19, the equalities hold if and only if

$$\psi \circ f \circ \varphi(z) = e^{i\tau} z = \rho_\tau(z).$$

Hence $f = \psi^{-1} \circ \rho_\tau \circ \varphi^{-1} \in \mathrm{Aut}(D)$, and this concludes the proof.

In fact, we can define a metric d on D as

$$d_z s^2 = \frac{|dz|^2}{(1 - |z|^2)^2}.$$

(For hyperbolic metric, Poincaré metric, cf. Chapter 5, Section 5.1.) Then (2.29) becomes $d_w s^2 \leq d_z s^2$. Therefore, if $w = f(z)$ is holomorphic on D and maps D into D, then the Poincaré metric does not increase under f. Poincaré metric does not change under f if and only if $f \in \mathrm{Aut}(D)$. This is another statement of Theorem 2.21 and it gives a clear geometric explanation of Schwarz Lemma.

2.6 Integral Representation of Holomorphic Functions

Cauchy Integral Formula (2.8) is an integral representation of holomorphic functions. Define

$$H(\zeta, z) = \frac{1}{2\pi i}\frac{1}{\zeta - z},$$

and call it the *Cauchy kernel*. Then the equation (2.8) can be written as

$$f(z) = \int_{\partial U} f(\zeta)H(\zeta, z)\, d\zeta.$$

In other words, the value of a holomorphic function $f(z)$ at a point $z \in U$ can be represented as the integral of Cauchy kernel $H(\zeta, z)$, and $f(\zeta)$, the value of f on ∂U.

From this, some other integral representations can also be derived. Let's look at the integral representation of harmonic functions first.

Assume that $U(z)$ is a harmonic function on the unit disc D and it is continuous on $\bar{D}$. Then by the mean-value property of harmonic function we have

$$\frac{1}{2\pi}\int_0^{2\pi} U(e^{i\psi})\,d\psi = U(0). \tag{2.30}$$

If $a \in D$, then $w = (z-a)/(1-\bar{a}z) \in \operatorname{Aut}(D)$ and it maps ∂D to ∂D. Let $U(w) = u(z)$. Then $u(z)$ is also a harmonic function on D and $u(0) = u(a)$. If $z = e^{i\tau}$ is mapped to $w = e^{i\psi}$, or equivalently $e^{i\psi} = (e^{i\tau} - a)/(1 - \bar{a}e^{i\tau})$, then

$$d\psi = \frac{1-|a|^2}{|1-\bar{a}e^{i\tau}|^2}\,d\tau.$$

Substitution in (2.30) shows that

$$\frac{1}{2\pi}\int_0^{2\pi} u(e^{i\tau})\frac{1-|a|^2}{|1-\bar{a}e^{i\tau}|^2}\,d\tau = u(a). \tag{2.31}$$

Denote

$$P(\zeta, a) = \frac{1}{2\pi}\frac{1-|a|^2}{|1-\bar{a}e^{i\tau}|^2} = \frac{1}{2\pi}\frac{1-|a|^2}{|\zeta - a|^2},$$

where $\zeta = e^{i\tau}$. This is called the *Poisson kernel*. By changing a to z, (2.31) becomes

$$\int_0^{2\pi} u(\zeta)P(\zeta, z)\,d\tau = u(z).$$

The equation (2.31) is called the *Poisson integral formula*, it is a integral representation of a function which is harmonic in the unit disc D and continuous on $\bar{D}$. In other words, the value of a harmonic function at a point inside the unit disc can be represented as the integral of the Poisson kernel and the value of the harmonic function on the unit circle ∂D.

The equation (2.31) can be generalized. If a function u is harmonic on $D(0, R)$ and continuous on $\bar{D}(0, R)$, then

$$\frac{1}{2\pi}\int_0^{2\pi} u(\zeta)\frac{R^2 - |z|^2}{|\zeta - z|^2}\,d\tau = u(z), \tag{2.32}$$

where $\zeta = Re^{i\tau}$ and $z \in D(0, R)$. Let

$$P(\zeta, z) = \frac{1}{2\pi}\frac{R^2 - |z|^2}{|\zeta - z|^2}.$$

This is also referred to as the Poisson kernel. It is easy to see that (2.32) is equivalent to

$$u(z) = \mathrm{Re}\left[\frac{1}{2\pi i}\int_{|\zeta|=R} u(\zeta)\frac{\zeta + z}{\zeta - z}\frac{d\zeta}{\zeta}\right]. \tag{2.33}$$

The function inside of the bracket of the above equation is holomorphic in $|z| < R$. Hence, u is the real part of the holomorphic function

$$f(z) = \frac{1}{2\pi i}\int_{|\zeta|=R} u(\zeta)\frac{\zeta + z}{\zeta - z}\frac{d\zeta}{\zeta} + ic, \tag{2.34}$$

where c is an arbitrary constant. Thus, $f(z) = u(z) + iv(z)$.

The imaginary part of (2.34) is

$$v(z) = \frac{1}{2\pi}\int_0^{2\pi} u(\zeta)\frac{2\,\mathrm{Im}(z\bar{\zeta})}{|\zeta - z|^2}\,d\tau + c, \tag{2.35}$$

where $\zeta = Re^{i\tau}$. Obviously we have $c = v(0)$.

Let

$$S(\zeta, z) = \frac{1}{2\pi i}\frac{\zeta + z}{\zeta - z}\frac{1}{\zeta}.$$

This is called the *Schwarz kernel*. Then (2.34) becomes

$$f(z) = \int_{|\zeta|=R} u(\zeta)S(\zeta, z)\,d\zeta + iv(0).$$

This is another integral representation of holomorphic functions. In other words, the value of holomorphic function f at a point z in $D(0, R)$ can be represented as the integral of Schwarz kernel $S(\zeta, z)$ and the real part of f on $\partial D(0, R)$.

Similarly, let

$$Q(\zeta, z) = \frac{1}{\pi} \frac{\operatorname{Im}(z\bar{\zeta})}{|\zeta - z|}.$$

It is called the *conjugate Poisson kernel*. Then equation (2.35) becomes

$$v(z) = \int_0^{2\pi} u(\zeta) Q(\zeta, z)\, d\zeta + v(0).$$

This is another integral representation of harmonic functions. In other words, the value of a harmonic function v on a point z in $D(0, R)$ can be represented as the integral of Poisson kernel $Q(\zeta, z)$ and $u(\zeta)$, the value of the conjugate of v on $\partial D(0, R)$. Two harmonic functions are said to be *conjugate* to each other if they are the real and the imaginary parts of a holomorphic function respectively.

Because of the connections between the Poisson kernel and other branches of mathematics, the Poisson kernel is also important besides the Cauchy kernel. We illustrate two of the connections here, in partial differential equations and in harmonic analysis.

Poisson Integral and Partial Differential Equations

An important problem in the theory of partial differential equations is to solve elliptic equations under Dirichlet conditions. That is, to find a function that satisfies a given elliptic equation in a region and is equal to a given function on the boundary of the region.

To find a function that satisfies the Laplace equation in $D(0, R)$ and is equal to a given continuous function $\varphi(Re^{i\tau})$ on $\partial D(0, R)$, is a Dirichlet problem which can be solved by the Poisson integral formula (2.32). The solution of this problem is

$$u(z) = \int_0^{2\pi} P(\zeta, z)\varphi(\zeta)\, d\tau,$$

where $\zeta = Re^{i\tau}$ and this solution is unique.

Indeed, since

$$P(\zeta, z) = \frac{1}{2\pi} \frac{|\zeta|^2 - |z|^2}{|\zeta - z|^2} = \frac{1}{2\pi} \left(\frac{\zeta}{\zeta - z} + \frac{\bar{z}}{\bar{\zeta} - \bar{z}} \right),$$

we have $\Delta P = 0$, where $\Delta = 4(\partial/\partial z)(\partial/\partial \bar{z})$. Obviously,

$$u(z) = \int_0^{2\pi} P(\zeta, z)\varphi(\zeta)\, d\tau$$

satisfies $\Delta u(z) = 0$ on $D(0, R)$. We want to show that $\lim_{z\to\xi} u(z) = \varphi(\xi)$, for $z \in D(0, R)$ and $\xi \in \partial D(0, R)$. Since

$$\int_0^{2\pi} P(\zeta, z)\, d\zeta = 1,$$

we have

$$u(\rho e^{i\theta}) - \varphi(Re^{i\theta_0}) = \int_0^{2\pi} P(\zeta, z)(\varphi(\zeta) - \varphi(Re^{i\theta_0}))\, d\tau,$$

where $\xi = Re^{i\theta_0}$, $z = \rho e^{i\theta}$ and $0 < \rho < R$.

By the assumption that φ is a continuous function on $|\zeta| = R$, for any $\varepsilon > 0$, there exists a $\delta > 0$ such that $|\varphi(Re^{i\theta_0}) - \varphi(Re^{i\tau})| < \varepsilon$ when $|\theta_0 - \tau| < \delta$. Thus

$$\begin{aligned} &u(\rho e^{i\theta}) - \varphi(Re^{i\theta_0}) \\ &\quad = \left(\int_{|\theta_0-\tau|<\delta} + \int_{|\theta_0-\tau|\geq\delta}\right) P(\zeta, z)(\varphi(\zeta) - \varphi(Re^{i\theta_0}))\, d\tau \\ &\quad = I_1 + I_2. \end{aligned}$$

It is easy to see that

$$|I_1| < \varepsilon \int_{|\theta_0-\tau|<\delta} P(\zeta, z)\, d\tau < \varepsilon.$$

Since

$$P(\zeta, z) = \frac{1}{2\pi} \frac{|\zeta|^2 - |z|^2}{|\zeta - z|^2}$$

and $|\theta_0 - \tau| > \delta$ in I_2, for any given $\varepsilon > 0$ and $\delta > 0$, there exists a $\eta > 0$ such that $|\zeta - z| > 2R^2(1 - \cos\delta)$ and $R^2 - \rho^2 < (R^2/M)(1 - \cos\delta)\varepsilon$ when $|z - \xi| < \eta$ and $|\theta_0 - \tau| > \delta$, where $M = \sup_{|\zeta|=R} |\varphi(\zeta)|$. Thus $|I_2| < \varepsilon$. Therefore, $|u(\rho e^{i\theta}) - \varphi(Re^{i\theta_0})| < 2\varepsilon$ as ρ is sufficiently close to R and the equation $\lim_{\rho\to R} u(\rho e^{i\theta}) = \varphi(Re^{i\theta_0})$ follows.

It is not difficult to prove the uniqueness. Suppose the contrary. Let u and v be two different solutions. Then u and v have the same value $\varphi(Re^{i\theta})$ on the boundary of $D(0, R)$. It follows that $u - v$ vanishes on $\partial D(0, R)$. Since $u - v$ is also a harmonic function, by Poisson integral formula, the only harmonic function vanishes on the boundary of a region is the zero function.

Poisson Integral and Harmonic Analysis

Take $R = 1$ and consider the unit disc D in (2.32), then the left hand side of (2.32) becomes

$$\frac{1}{2\pi}\int_0^{2\pi} u(\zeta)\frac{1-|z|^2}{|\zeta - z|^2}\,d\tau = \frac{1}{2\pi}\int_0^{2\pi} \frac{(1-\rho^2)u(e^{i\tau})d\tau}{1-2\rho\cos(\theta-\tau)+\rho^2} \tag{2.36}$$

where $z = \rho e^{i\theta}$ and $\zeta = e^{i\tau}$.

If $u(e^{i\tau})$ is a given continuous function on the unit circle ∂D, then $u(e^{i\tau})$ has the Fourier series

$$u(e^{i\tau}) \sim \sum_{n=-\infty}^{\infty} a_n e^{in\tau}, \quad a_n = \frac{1}{2\pi}\int_{-\pi}^{\pi} u(e^{i\tau})e^{-in\tau}\,d\tau.$$

This Fourier series is not necessary convergent, but we can form its Abel sum

$$\sum_{n=-\infty}^{\infty} a_n \rho^{|n|} e^{in\tau}.$$

By a simple calculation, we can find that this Abel sum is just the integral on the right hand side of (2.36)

$$\frac{1}{2\pi}\int_0^{2\pi} \frac{(1-\rho^2)u(e^{i\tau})d\tau}{1-2\rho\cos(\theta-\tau)+\rho^2}.$$

By the solution of Dirichlet problem we discussed before, this integral tends to $u(e^{i\theta})$ when $\rho \to 1$,

$$\lim_{\rho\to 1}\sum_{n=-\infty}^{\infty} a_n \rho^{|n|} e^{in\tau} = u(e^{i\tau}).$$

In other words, for the continuous function $u(e^{i\tau})$ on the unit circle ∂D, the Abel sum of its Fourier series tends to $u(e^{i\tau})$ itself as $\rho \to 1$. The Fourier series of a continuous function is Abel summable. This is a elementary but important theorem in harmonic analysis.

Finally we use the solution of Dirichlet problem to prove that a continuous function satisfying the mean-value property is a harmonic function. This assertion was mentioned in Section 2.5. Indeed, we can show that

Theorem 2.22 *If $f(z)$ is a real valued continuous function on a region $U \subseteq \mathbb{C}$ and $f(z)$ satisfies the local mean-value property in U, that is, for*

any point $z_0 \in U$, there exists an $r_0 > 0$ sufficiently small such that

$$\frac{1}{2\pi}\int_0^{2\pi} f(z_0 + re^{i\theta})\, d\theta = f(z_0) \tag{2.37}$$

when $0 < r < r_0$, then $f(z)$ is harmonic on U.

Proof For an arbitrary point $z_0 \in U$, there exists an $r_0 > 0$ such that (2.37) is true for all r with $0 < r \leq r_0$. Let $u_0(\theta) = f(z_0 + r_0e^{i\theta})$. Then we can find a function $u(z)$ that is harmonic in $D(z_0, r_0)$ and has value $u_0(\theta)$ on the boundary of U by solving the Dirichlet problem. Consider the function $f(z) - u(z)$ on $\bar{D}(z_0, r_0)$. Since both $f(z)$ and $u(z)$ satisfy the mean value property, so does $f(z) - u(z)$. By the discussion in Section 2.5, we know that Maximum Modulus Principle can be derived from Mean Value Property. Thus the maximum value of $|f(z) - u(z)|$ is assumed on $\partial D(z_0, r_0)$. Since $f(z) - u(z)$ is zero on $\partial D(z_0, r_0)$, by the Maximum Modulus Principle, $|f(z) - u(z)| \leq 0$. Since $f(z)$ and $u(z)$ are continuous, we have $f(z) = u(z)$ when $z \in D(z_0, r_0)$. It follows that $f(z)$ is harmonic at $z = z_0$. Hence $f(z)$ is harmonic on U since z_0 is arbitrary on U.

Exercise II

1. Calculate the following by using the Cauchy Integral Formula.

(i) $\displaystyle\int_{|z+i|=3} \sin z \frac{dz}{z+i}$; (ii) $\displaystyle\int_{|z|=2} \frac{e^z}{z-1}\, dz$;

(iii) $\displaystyle\int_{|z|=4} \frac{\cos z}{z^2 - \pi^2}\, dz$;

(iv) $\displaystyle\int_{|z|=2} \frac{dz}{(z-1)^n(z-3)}$, $n = 1, 2, \cdots$;

(v) $\displaystyle\int_{|z|=\frac{3}{2}} \frac{dz}{(z^2+1)(z^2+4)}$; (vi) $\displaystyle\int_{|z|=2} \frac{dz}{z^5 - 1}$;

(vii) $\displaystyle\int_{|z|=R} \frac{dz}{(z-a)^n(z-b)}$,
where a, b are not on the circle $|z| = R$ and n is a positive integer.

(viii) $\displaystyle\int_{|z|=3} \frac{dz}{(z^3-1)(z-2)^2}$.

2. Prove that

$$\left(\frac{z^n}{n!}\right)^2 = \frac{1}{2\pi i}\int_C \frac{z^n e^{z\zeta}}{n!\zeta^n}\cdot\frac{d\zeta}{\zeta},$$

where C is a simple closed curve around the origin.

3. Let f and g be holomorphic in the unit disc $|z|<1$ and continuous on $|z|\leq 1$. Show that

$$\frac{1}{2\pi i}\int_{|\zeta|=1}\left(\frac{f(\zeta)}{\zeta - z} + \frac{zg(\zeta)}{z\zeta - 1}\right)d\zeta = \begin{cases} f(z), & |z|<1,\\ g\left(\frac{1}{z}\right), & |z|>1.\end{cases}$$

4. Let

$$f(z) = \int_{|\zeta|=3}\frac{3\zeta^2+7\zeta+1}{\zeta - z}\,d\zeta.$$

Find $f'(1+i)$.

5. Show that

$$\int_0^{2\pi}\cos^{2n}\theta d\theta = 2\pi\cdot\frac{1\cdot 3\cdot 5\cdot\cdots\cdot(2n-1)}{2\cdot 4\cdot 6\cdot\cdots\cdot 2n}$$

by calculating

$$\int_{|z|=1}\left(z+\frac{1}{z}\right)^{2n}\frac{dz}{z}.$$

6. Let

$$p_n(z) = \frac{1}{2^n n!}\frac{d^n}{dz^n}[(z^2-1)^n].$$

This is referred to as *Legendre polynomial.* Show that

$$p_n(z) = \frac{1}{2\pi i}\int_\gamma \frac{(\zeta^2-1)^n}{2^n(\zeta - z)^{n+1}}\,d\zeta,$$

where γ is a simple closed curve containing z.

7. Let $f(z)$ be a holomorphic function on $\mathbb{C}$ and $|f(z)|\leq Me^{|z|}$. Show that $|f(0)|\leq M$ and

$$\frac{|f^n(0)|}{n!}\leq M\left(\frac{e}{n}\right)^n,\quad n=1,2,\cdots.$$

8. (*Cauchy integral formula for the outside region of a simple closed curve*) Let γ be a rectifiable simple closed curve, D_1 be the inside of γ

and D_2 be the outside of γ. Suppose a function $f(z)$ is holomorphic in D_2 and continuous on $D_2 \cup \gamma$. Show that

(i) If $\lim_{z\to\infty} f(z) = A$, then

$$\frac{1}{2\pi i}\int_\gamma \frac{f(\zeta)}{\zeta - z}\,d\zeta = \begin{cases} -f(z) + A, & z \in D_2, \\ A, & z \in D_1; \end{cases}$$

(ii) If the origin is in D_1, then

$$\frac{1}{2\pi i}\int_\gamma \frac{zf(\zeta)}{z\zeta - \zeta^2}\,d\zeta = \begin{cases} f(z), & z \in D_2, \\ 0, & z \in D_1. \end{cases}$$

9. Let $f(z)$ be holomorphic in a bounded region D and continuous on $\bar{D}$ with $f(z) \neq 0$. Show that if $|f(z)| = M$ on ∂D, then $f(z) = Me^{i\alpha}$, where α is a real number.

10. Suppose $f(z)$ is holomorphic and bounded in $\mathbb{C}$, a, b are complex numbers. Find

$$\lim_{R\to\infty}\int_{|z|=R} \frac{f(z)}{(z-a)(z-b)}\,dz.$$

This leads to another proof of Liouville Theorem.

11. If $f(z)$ is holomorphic in the unit disc $|z| < 1$ and

$$|f(z)| \leq \frac{1}{1-|z|} \quad (|z| < 1),$$

then

$$|f^{(n)}(0)| \leq \frac{n!}{r^n(1-r)}$$

for $0 < r < 1$. Especially, if $r = 1 - 1/(n+1)$, then

$$|f^{(n)}(0)| < e(n+1)!, \quad n = 1, 2, \cdots.$$

12. (*Integral of Cauchy type*) If the function $\varphi(\zeta)$ is continuous on a rectifiable curve γ, show that the function

$$\Phi(z) = \frac{1}{2\pi i}\int_\gamma \frac{f(\zeta)d\zeta}{\zeta - z}$$

is holomorphic in any region D which does not contain any point of γ. Moreover,

$$\Phi^{(n)}(z) = \frac{n!}{2\pi i}\int_\gamma \frac{\varphi(\zeta)d\zeta}{(\zeta - z)^{n+1}}, \quad n = 1, 2, \cdots.$$

13. Suppose that a non-constant function $f(z)$ is holomorphic on a bounded region D and continuous on $\bar{D}$. Let $m = \inf_{\tau \in \partial D} |f(z)|$, $M = \sup_{z \in \partial D} |f(z)|$ and $f(z) \neq 0$. Then $m < |f(z)| < M$ for any point $z \in D$.

14. If $p_n(z)$ is a polynomial with degree n and $|p_n(z)| \leq M$ for $|z| < 1$, then $|p_n(z)| \leq M|z|^n$ for $1 \leq |z| < \infty$.

15. Suppose $f(z)$ is a holomorphic function on the disc $|z| < R$, $|f(z)| \leq M$ and $f(0) = 0$. Show that

$$|f(z)| \leq \frac{M}{R}|z|, \quad |f'(0)| \leq \frac{M}{R},$$

where the equalities hold if and only if $f(z) = (M/R)e^{i\alpha}z$ with $\alpha \in \mathbb{R}$.

16. Prove Corollary 2.1 by applying Theorem 2.7.

17. Prove that if $\{f_n(z)\}$ is a sequence of continuous functions on a region D that converges uniformly to $f(z)$, then

$$\int_\gamma f(z)\,dz = \lim_{n\to\infty} \int_\gamma f_n(z)\,dz$$

for any curve γ in D. (This assertion was used in the proof of Theorem 2.15.)

18. Let $f(z)$ be a holomorphic function in the disc $|z| < 1$ with $\operatorname{Re} f(z) > 0$ and $f(0) = \alpha > 0$. Show that

$$\left|\frac{f(z) - \alpha}{f(z) + \alpha}\right| \leq |z|, \quad |f'(0)| \leq 2\alpha.$$

19. Let $f(z)$ be a holomorphic function in the disc $|z| < 1$ with $f(0) = 0$ and $\operatorname{Re} f(z) \leq A\,(A > 0)$. Show that

$$|f(z)| \leq \frac{2A|z|}{1 - |z|}.$$

20. Find the Taylor expansions and their radius of convergence of the following functions at $z = 0$.

(i) $\dfrac{e^z + e^{-z} + 2\cos z}{4}$; (ii) $\dfrac{z^2 + 4z^4 + z^6}{(1 - z^2)^3}$;

(iii) $(1 - z^{-5})^{-4}$; (iv) $\dfrac{z^6}{(z^2 - 1)(z + 1)}$.

21. (1) If $f(z)$ is holomorphic on $|z| \leq r$ and

$$f(z) = \sum_{n=0}^{\infty} a_n z^n$$

is its Taylor expansion at $z = 0$, then

$$\sum_{n=0}^{\infty} |a_n|^2 r^{2n} = \frac{1}{2\pi} \int_{-\pi}^{\pi} |f(re^{i\theta})|^2 \, d\theta.$$

(2) If $r = 1$ in (1), then

$$\sum_{n=0}^{\infty} \frac{|a_n|^2}{n+1} = \frac{1}{\pi} \iint_D |f(z)|^2 \, dA,$$

where D is the unit disc $|z| \le 1$ and dA is the area element.

22. Verify the equations (2.32) and (2.33).

23. Show that one of the zeros of the equation $z^4 - 6z + 3 = 0$ is in the disc $|z| < 1$ and the other three zeros are in $1 < |z| < 2$.

24. Find the number of zeros of the equation $z^7 - 5z^4 - z + 2 = 0$ in the disc $|z| < 1$.

25. Show that there is exactly one zero of the equation $z^4 + 2z^3 - 2z + 10 = 0$ in each quadrant.

26. Find the number of zeros of the equation $z^4 - 8z + 10 = 0$ in the disc $|z| < 1$ and in the annulus $1 < |z| < 3$.

27. Show that if $a > e$, then there are n zeros in the disc $|z| < 1$ for the equation $e^z = az^n$.

28. If $f(z)$ is holomorphic in the disc $|z| < 1$ and continuous on $|z| \le 1$ and $|f(z)| < 1$, then there is a unique fixed point of f in the disc $|z| < 1$.

29. (1) Find a holomorphic function whose real part is

$$e^x (x \cos y - y \sin y),$$

where $x + iy = z$.

(2) Find the most general harmonic function of the form

$$ax^3 + bx^2 y + cxy^2 + dy^3$$

where a, b, c, d are real numbers.

30. By applying the mean-value property of harmonic functions, show that if $-1 < r < 1$, then

$$\int_0^{\pi} \ln(1 - 2r\cos\theta + r^2) \, d\theta = 0.$$

31. Prove that if a harmonic function $u(z)$ is bounded on the entire complex plane $\mathbb{C}$, then $u(z)$ is equal to a constant identically.

32. Find a harmonic function on $|z| < 1$ such that its value is 1 on an arc of $|z| = 1$ and zero on the rest of $|z| = 1$.

33. Suppose U is a region, $f_i(z)\,(i = 1, 2, \cdots)$ are holomorphic on U and continuous on $\bar{U}$. Show that if

$$\sum_{n=1}^{\infty} f_n(z)$$

converges uniformly on the boundary of U, then it converges uniformly on $\bar{U}$.

34. Suppose $f(z)$ is holomorphic on $D(0, R)$ and continuous on $\overline{D(0, R)}$. Let $M = \max_{|z|=R} |f(z)|$. Show that if $z_0 \in D(0, R) \setminus \{0\}$ is a zero of $f(z)$, then

$$R|f(0)| \leq (M + |f(0)|)|z_0|.$$

35. Let $|z_1| > 1$, $|z_2| > 1$, $\cdots$, $|z_n| > 1$. show that there exists a point z_0 such that $\prod_{k=1}^{n} |z_0 - z_k| > 1$.

36. Suppose that $f(z)$ is holomorphic on $D(0, R)$. Show that

$$M(r) = \max_{|z|=r} |f(z)|$$

is a increasing function on $[0, R)$.

37. Use the Maximum Modulus Principle to prove the Fundamental Theorem of Algebra.

38. If $f(z)$ is a non-constant holomorphic function on a region U and has no zeros in U, then $|f(z)|$ can not attain its minimal value in U.

39. (*Hadamard's Three Circles Theorem*) Suppose that $0 < r_1 < r_2 < \infty$, $U = \{z \in \mathbb{C} \,|\, r_1 < |z| < r_2\}$, $f(z)$ is holomorphic on U and continuous on $\bar{U}$. Define $M(r) = \max_{|z|=r} |f(z)|$. Show that $\log M(r)$ is a convex function of $\log r$ on $[r_1, r_2]$. In other words, the inequality

$$\log M(r) \leq \frac{\log r_2 - \log r}{\log r_2 - \log r_1} \log M(r_1) + \frac{\log r - \log r_1}{\log r_2 - \log r_1} \log M(r-2)$$

holds for $r \in [r_1, r_2]$.

40. If $f(z)$ is holomorphic on $D(0, 1)$ and $f(0) = 0$, show that

$$\sum_{n=1}^{\infty} f(z^n)$$

converges absolutely on $D(0, 1)$ and converges uniformly on every compact subset of $D(0, 1)$.

41. Let $f(z)$ be a holomorphic function on $D(0,R)$ and $f(0)=0$. If $f(D(0,R)) \subset D(0,M)$, then

(i) the inequalities

$$|f(z)| \le \frac{M}{R}|z|, \quad |f'(0)| \le \frac{M}{R}$$

hold for $z \in D(0,R) \setminus \{0\}$.

(ii) the equalities in (1) hold if and only if

$$f(z) = \frac{M}{R}e^{i\theta}z,$$

where θ is a real number.

42. Let $f(z)$ be a holomorphic function on $D(0,1)$ and $f(0)=0$. If there exists a constant $A>0$ such that $\operatorname{Re} f(z) \le A$ for $z \in D(0,1)$, then

$$|f(z)| \le \frac{2A|z|}{1-|z|}$$

for $z \in D(0,1)$.

43. Let $f(z)$ be a holomorphic function on $D(0,1)$ and $f(0)=1$. If $\operatorname{Re} f(z) \ge 0$ for every $z \in D(0,1)$, using Schwarz Lemma, show that

(1) the inequality

$$\frac{1-|z|}{1+|z|} \le \operatorname{Re} f(z) \le |f(z)| \le \frac{1+|z|}{1-|z|}$$

holds for $z \in D(0,1)$;

(2) the equality in (1) holds for $z \ne 0$ if and only if

$$f(z) = \frac{1+e^{i\theta}z}{1-e^{i\theta}z},$$

where θ is an arbitrary real number.

44. Suppose $f(z)$ is holomorphic on $D(0,1)$. Show that there exists a sequence $\{z_n\}$ in $D(0,1)$ that converges to a point $z_0 \in \partial D(0,1)$ such that $\lim_{n\to\infty} f(z_n)$ exists.

45. Let $f(z)$ be a holomorphic function on $D(0,1)$ and $f(D(0,1)) \subset D(0,1)$. If $z_1, z_2, \cdots, z_n$ are all of the different zeros of $f(z)$ in $D(0,1)$ and their multiplicities are $k_1, k_2, \cdots, k_n$ respectively, then

$$|f(z)| \le \prod_{j=1}^{n} \left| \frac{z-z_j}{1-\bar{z}_j z} \right|^{k_j}$$

for all $z \in D(0,1)$. Especially

$$|f(0)| \le \prod_{j=1}^{n} |z_j|^{k_j}.$$

46. If $f(z)$ is holomorphic on $D(0,1)$ and $f(D(0,1)) \subset D(0,1)$, then

$$M|f'(0)| \le M^2 - |f(0)|^2.$$

47. Let $f(z)$ be a holomorphic function on $D(0,1)$ and $f(0) = 0$. If $|\operatorname{Re} f(z)| < 1$ for all $z \in D(0,1)$, then

$$|\operatorname{Re} f(z)| \le \frac{4}{\pi} \arctan |z|, \quad |\operatorname{Im} f(z)| \le \frac{2}{\pi} \log \left(\frac{1+|z|}{1-|z|} \right)$$

for all $z \in D(0,1)$.

48. Find the group of holomorphic automorphisms of the upper half plane $\mathbb{C}^+ = \{z \in \mathbb{C} |\ \operatorname{Im} z > 0\}$, $\operatorname{Aut}(\mathbb{C}^+)$.

49. If $f(z)$ is holomorphic on $D(0,1) \cup \{1\}$, $f(0) = 0$, $f(1) = 1$ and $f(D(0,1)) \subset D(0,1)$, then $f'(1) \ge 1$.

50. If $f(z)$ is a bounded entire function and z_1, z_2 are arbitrary points in $D(0,r)$, then

$$\int_{|z|=r} \frac{f(z)}{(z-z_1)(z-z_2)}\, dz = 0,$$

and this implies the Liouville Theorem.

Appendix I Partition of Unity

On the complex plane $\mathbb{C}$, define

$$\theta(z) = \begin{cases} k \exp\left\{ \dfrac{-1}{1-|z|^2} \right\}, & |z| < 1, \\ 0, & |z| \ge 1, \end{cases}$$

where k is a constant such that

$$\int_{\mathbb{C}} \theta(z)\, dA = 1,$$

and dA is an area element. It is easy to see that $\theta(z)$ is a C^∞ function on $\mathbb{C}$ and strictly greater than zero in $|z| < 1$. The support of $\theta(z)$, $\operatorname{supp} \theta(z)$, is $|z| \le 1$. The function $\theta(z)$ is called a *standard function* on $\mathbb{C}$. For $\varepsilon > 0$, if $\theta_\varepsilon(z) = \varepsilon^{-2}\theta(z/\varepsilon)$, then θ_ε has the same properties as θ and $\operatorname{supp}\theta_\varepsilon(z)$ is $|z| \leq \varepsilon$.

Let $\Omega \subset \mathbb{C}$ be an open set, $\mathfrak{D}(\Omega)$ be the space of all C^∞ real functions on $\mathbb{C}$ whose compact support are in Ω. Then we have

Theorem 2.23 *If $\Omega \subset \mathbb{C}$ is an open set, $\mathfrak{B}$ is an open basis of Ω, then there exists a sequence U_1, U_2, $\cdots$, U_n, $\cdots$ in $\mathfrak{B}$, such that*

(1) $\bigcup\limits_{j\geq 1} U_j = \Omega$;

(2) any compact subset K of Ω intersects with only finitely many elements in $\{U_j\}_{j\geq 1}$.

Proof Suppose $K_{-1} = \emptyset$, $K_0 = \emptyset$, $K_1, K_2, \cdots, K_n, \cdots$ is a sequence of compact subsets of Ω which exhausts Ω. That is, the sequence satisfies

(i) K_j is contained in $\overset{\circ}{K}_{j+1}$, the interior of K_{j+1}

(ii) $\Omega = \bigcup_{j\geq 1} K_j$.

Let $W_r = \overset{\circ}{K}_{r+1} \setminus K_{r-2}$ and $V_r = K_r \setminus \overset{\circ}{K}_{r-1}$ for $r \geq 1$. Then W_r is open, V_r is compact, $V_r \subseteq W_r$ and $\Omega = \bigcup_{r>1} V_r$.

For every point z in V_r, there exists an $U_{z,r} \in \mathfrak{B}$ such that $z \in U_{z,r} \subseteq W_r$. Since V_r is compact, there exist finite many points $z_{r,1}, \cdots, z_{r,k}$ in V_r such that

$$V_r \subseteq \bigcup_{1\leq i\leq k_i} U_{z_{r,i},r} \subseteq W_r.$$

The sequence $\{U_{z_{r,i},r}\}_{r\geq 1}$ is countable and satisfies (1) and (2) since any compact subset K of Ω intersects only finite many elements in $\{W_r\}$.

The sequence $\{U_j\}_{j\geq 1}$ which satisfies (1) is called an *open covering* of Ω. Property (2) means that the open covering is *locally finite.*

Theorem 2.24 ***(Partition of Unity Theorem)*** *If Ω is a non-empty open subset of $\mathbb{C}$, $\{\Omega_i\}_{i\in I}$ is an open covering of Ω where I is the set of non-negative integers, then there exists a sequence $\alpha_1(z), \alpha_2(z), \cdots, \alpha_n(z), \cdots$ in $\mathfrak{D}(\Omega)$, such that*

(1) for every $j \geq 1$, there exists an $i = i(j) \in I$ such that $\operatorname{supp}\alpha_j \subseteq \Omega_i$ and $\{\operatorname{supp}\alpha_j\}_{j\geq 1}$ is locally finite;

(2) $0 \leq \alpha_j \leq 1$ for every $j \geq 1$;

(3) $\sum_{j\geq 1}\alpha_j(z) = 1$ for every $z \in \Omega$.

The sequence $\{\alpha_j(z)\}_{j\geq 1}$ is called a C^∞ *partition of unity* with respect to the open covering $\{\Omega_i\}_{i\in I}$.

Proof of Theorem 2.24 For every $z \in \Omega$, there exists a $r_z > 0$, such that $\overline{B(z, r_z)} \subseteq \Omega_{i_z}$, where $i_z \in I$ and $B(z, r_z)$ is a disc with center z and radius r_z. All of the $B(z,r)$ $(0 < r < r_z)$ form an open basis of Ω. Thus, by the proof of the last theorem, there exists a sequence $\{B(z_j, r_j)\}_{j\geq 1}$ that satisfies conditions (1) and (2) in the last theorem, and

$$B(z_j, r_j) \subseteq \overline{B(z_j, r_j)} \subseteq \Omega_{i(j)},$$

where $i(j) = i_{z_j}$. Suppose θ is a standard function. Let $\beta_j(z) = \theta_{r_j}(z - z_j)$. Then $\beta_j \in \mathfrak{D}(\Omega)$ and $\{\operatorname{supp} \beta_j\}_{j\geq 1}$ is locally finite. Thus

$$s(z) = \sum_{j\geq 1} \beta_j(z)$$

is a C^∞ function on Ω and $s(z) > 0$ for $z \in \Omega$. Let $\alpha_j(z) = \beta_j(z)/s(z)$. Then the sequence $\{\alpha_j(z)\}$ has the required property.

Theorem 2.25 *Let $\Omega \subset \mathbb{C}$ be an open set, K be a compact subset of Ω, V be an open neighborhood of K and $V \subseteq \Omega$. Then there exists $\varphi \in \mathfrak{D}(V)$ such that*

(1) $0 \leq \varphi \leq 1$;

(2) $\varphi \equiv 1$ *on a neighborhood of* K.

Proof For any $\varepsilon > 0$, let $V(K, \varepsilon) = \{z \in \mathbb{C} |\ \operatorname{dist}(z, K) < \varepsilon\}$, where $\operatorname{dist}(z, K)$ is the distance between z and K. Choose an $\varepsilon > 0$ such that $K \subseteq V(K, \varepsilon) \subseteq V(K, 2\varepsilon) \subseteq V$. Let $\Omega_1 = V(K, 2\varepsilon)$ and $\Omega_2 = \Omega \setminus V(K, \varepsilon)$. Then Ω_1 and Ω_2 form an open covering of Ω. By the Partition of Unity Theorem, there exists a sequence $\{\alpha_j(z)\}_{j\geq 1}$ which satisfies the conditions (1), (2) and (3) of the last theorem. Define

$$\varphi(z) = \sum_j{}' \alpha_j(z),$$

where the notation $\sum'$ means the summation of α_j that satisfies $\operatorname{supp} \alpha_j(z) \subseteq \Omega_2$. Obviously, $\varphi(z) \in \mathfrak{D}(\Omega)$ and $\operatorname{supp} \alpha_j(z) \subseteq V(K, 2\varepsilon)$. It is easy to see that $\varphi \equiv 1$ on a neighborhood of K since for those js which do not appear in the sum $\sum_j'$, $\operatorname{supp} \alpha_j \not\subseteq \Omega_1$. Thus, $\operatorname{supp} \alpha_j \subseteq \Omega_2$. It follows that $\alpha_j(z) = 0$ on $\overline{V(K, \varepsilon)}$. Therefore $\varphi(z) = \sum_{j\geq 1} \alpha_j(z) = 1$ for $z \in V(K, \varepsilon)$. This proves the theorem.

Chapter 3

Theory of Series of Weierstrass

3.1 Laurent Series

Studying properties of functions by series is part of the Weierstrass theory. In Chapters 1 and 2, we studied series of functions and power series expansions of holomorphic functions. Most of the results are similar to the corresponding materials in calculus. What is important are the results that are different from calculus. Theorem 1.14 (3) in Chapter 1 is one of them. The major difference between real and complex series theories is the Laurent series. It is a powerful tool for studying singularities of functions.

Before we introduce the Laurent series, let us look at the Weierstrass Theorem about function series.

Theorem 3.1 ***(Weierstrass Theorem)*** *Suppose $\{f_n(z)\}$ is a sequence of functions where each $f_n(z)$ is defined and holomorphic in a region $U \subseteq \mathbb{C}$. Assume that $\sum_{n=1}^{\infty} f_n(z)$ converges uniformly to $f(z)$ on every compact subset of U. Then $f(z)$ is holomorphic on U and for every $k \in \mathbb{N}$, $\sum_{n=1}^{\infty} f_n^{(k)}(z)$ converges uniformly to $f^{(k)}(z)$ on every compact subset of U.*

This is a profound result. The reader can compare it with the theorem of the derivative of function series in calculus.

Proof of Theorem 3.1 By Theorem 1.15 in Chapter 1, $f(z)$ is continuous on U. Suppose K is a circle in U and the interior of the circle is completely contained in U. Let γ be a rectifiable closed curve in K. Since $\sum_{n=1}^{\infty} f_n(z)$ converges uniformly on γ, by Theorem 1.15 in Chapter 1 and the assumption that $f_n(z)\,(n = 1, 2, \cdots)$ are holomorphic in the interior of K, we have

$$\int_\gamma f(z)\,dz = \sum_{n=1}^{\infty} \int_\gamma f_n(z)\,dz = 0.$$

Since

$$\int_\gamma f(z)\,dz = 0,$$

by the Morera Theorem (Theorem 2.9 in Section 2.3), we have $f(z)$ is holomorphic in K.

If $z_0 \in U$, $r > 0$ and $\bar{D}(z_0, r) \subset U$, then $\sum_{n=1}^{\infty} f_n(\zeta)$ converges uniformly to $f(\zeta)$ on $\partial D(z_0, r)$. If $z \in D(z_0, r/2)$, then by Corollary 2.1 in Section 2.3, we have

$$\sup_{z\in \overline{D(z_0,r/2)}} \left| \sum_{j=1}^{n} f_j^{(k)}(z) - f^{(k)}(z) \right| \le c_n \sup_{z\in D(z_0,r)} \left| \sum_{j=1}^{n} f_j(z) - f(z) \right|,$$

and the right hand side of this inequality tends to zero as $n \to \infty$.

Thus, $\sum_{j=1}^{\infty} f_j^{(k)}(z)$ converges uniformly to $f^{(k)}(z)$ on $D(z_0, r/2)$.

Let $\bar{d}$ be the closure of a bounded region $d \subset U$. Then for every point $z \in \bar{d}$, $\sum_{j=1}^{\infty} f_n^{(k)}(z)$ converges uniformly to $f^{(k)}(z)$ in a neighborhood of z. These neighborhoods form an open covering of $\bar{d}$. By the Heine-Borel Theorem, it has a finite subcovering of $\bar{d}$. Therefore, $\sum_{j=1}^{\infty} f_j^{(k)}(z)$ converges uniformly to $f^{(k)}(z)$ on $\bar{d}$ and the theorem follows.

By the Maximum Modulus Principle, if every function in the sequence $\{f_n(z)\}$ is holomorphic in a bounded region $D \subset \mathbb{C}$ and continuous on $\bar{D}$, the series $\sum_{n=1}^{\infty} f_n(z)$ converges uniformly on ∂D, then $\sum_{n=1}^{\infty} f_n(z)$ converges uniformly on $\bar{D}$. Hence, the condition "$\sum_{n=1}^{\infty} f_n(z)$ converges uniformly to $f(z)$ on every compact subset of U" can be changed to "$\sum_{n=1}^{\infty} f_n(z)$ converges uniformly to $f(z)$ on any closed curve in U" and Theorem 3.1 still holds.

Now, let us study the Laurent series.

Suppose $a \in \mathbb{C}$ and $c_n\,(n = 0, \pm 1, \pm 2, \cdots)$ are complex constants. A series of the form

$$\sum_{n=-\infty}^{\infty} c_n(z-a)^n \tag{3.1}$$

is called the *Laurent series* at the point a. There are two parts in Laurent series, one is the power series $\sum_{n=0}^{\infty} c_n(z-a)^n$ with non-negative exponents; the other is the power series $\sum_{n=1}^{\infty} c_{-n}(z-a)^{-n}$ with negative exponents. If both parts are convergent at a point $z = z_0$, then we say that the Laurent series is convergent at the point. If the radius of convergence of $\sum_{n=0}^{\infty} c_n(z-a)^n$ is $R(> 0)$, then the series converges absolutely in $|z - a| < R$ and uniformly on every compact subset of this disc. Thus, the sum of this series, $\varphi(z)$, is holomorphic on $|z-a|<R$.

Let $\zeta = 1/(z-a)$. Then

$$\sum_{n=1}^{\infty} c_{-n}(z-a)^{-n} = \sum_{n=1}^{\infty} c_{-n}\zeta^n.$$

Suppose the radius of convergence of the above series is $\lambda(>0)$. Then the series converges absolutely in $|\zeta|<\lambda$ and uniformly on every compact subset of this disc. Thus, the series $\sum_{n=1}^{\infty} c_{-n}(z-a)^{-n}$ converges absolutely in $r<|z-a|<\infty$ and uniformly on every compact subset of this region, where $r=1/\lambda$. The sum of the series, $\psi(z)$, is holomorphic on $r<|z-a|<\infty$. If $r>R$, then (3.1) diverges everywhere. If $r=R$, then (3.1) diverges everywhere except on $|z-a|=R$. There are different circumstances on $|z-a|=R$. For instance:

$$\sum_{\substack{n=-\infty \\ n\neq 0}}^{\infty} \frac{z^n}{n^2}$$

converges at every point of $|z|=1$,

$$\sum_{n=-\infty}^{\infty} z^n$$

diverges at every point of $|z|=1$,

$$\sum_{\substack{n=-\infty \\ n\neq 0}}^{\infty} \frac{z^n}{n}$$

converges on $|z|=1$ except at the point $z=1$. If $r<R$, then (3.1) converges absolutely on $r<|z-a|<R$ and uniformly on every compact subset of this annular region. The series (3.1) diverges on the outside of this region. The annular region $r<|z-z|<R$ is called the *annulus of convergence* of the series (3.1). By Theorem 3.1, series (3.1) converges to a holomorphic function in this annulus. The function $\varphi(z)$ is holomorphic in $|z-a|<R$, the function $\psi(z)$ is holomorphic in $r<|z-a|<\infty$ and the function $f(z)=\varphi(z)+\psi(z)$ is holomorphic in $r<|z-a|<R$. The series

$$\sum_{n=0}^{\infty} c_n(z-a)^n$$

is called the *holomorphic part* of (3.1). The series

$$\sum_{n=1}^{\infty} c_{-n}(z-a)^{-n}$$

is called the *principle part* (or *singular part*) of (3.1). The major characteristics of the function $f(z)$ are determined by the principle part.

In summary, if the annulus of convergence of the Laurent series (3.1) is $V = \{z | r < |z-a| < R\}$, then (3.1) converges absolutely on V and uniformly on every compact subset of V. The sum $f(z)$ is holomorphic on this annulus.

On the other hand, we have

Theorem 3.2 *If $f(z)$ is holomorphic on the annulus $V = \{z | r < |z - a| < R \ (0 \le r < R < \infty)\}$, then*

$$f(z) = \sum_{n=-\infty}^{\infty} c_n(z-a)^n \tag{3.2}$$

on V, where

$$c_n = \frac{1}{2\pi i}\int_{|\zeta-a|=\rho} \frac{f(\zeta)d\zeta}{(\zeta-a)^{n+1}}, \quad (r < \rho < R). \tag{3.3}$$

The series expansion (3.2) is unique. It is called the *Laurent expansion* or *Laurent series* of $f(z)$ on V.

Proof of Theorem 3.2 Since the integral (3.3) is independent of ρ $(r < \rho < R)$. If $r < \rho_1 < \rho_2 < R$, then

$$\int_{|\zeta-a|=\rho_1} \frac{f(\zeta)d\zeta}{(\zeta-a)^{n+1}} = \int_{|\zeta-a|=\rho_2} \frac{f(\zeta)d\zeta}{(\zeta-a)^{n+1}}.$$

Suppose $z \in V$, $\gamma_1 = \partial D(a, r_1)$, $\gamma_2 = \partial D(a, r_2)$, where $r < r_1 < r_2 < R$ and z is in the annulus $r_1 < |z-a| < r_2$ (see Figure 3). By the Cauchy Integral Formula, we have

$$f(z) = \frac{1}{2\pi i}\int_{\gamma_2} \frac{f(\zeta)d\zeta}{\zeta - z} - \frac{1}{2\pi i}\int_{\gamma_1} \frac{f(\zeta)d\zeta}{\zeta - z}. \tag{3.4}$$

If $\zeta \in \gamma_1$, then $|(\zeta-a)/(z-a)| < 1$. So we have

$$\frac{1}{\zeta - z} = \frac{-1}{(z-a)\left(1 - \dfrac{\zeta-a}{z-a}\right)} = -\sum_{n=1}^{\infty} \frac{(\zeta-a)^{n-1}}{(z-z)^n},$$

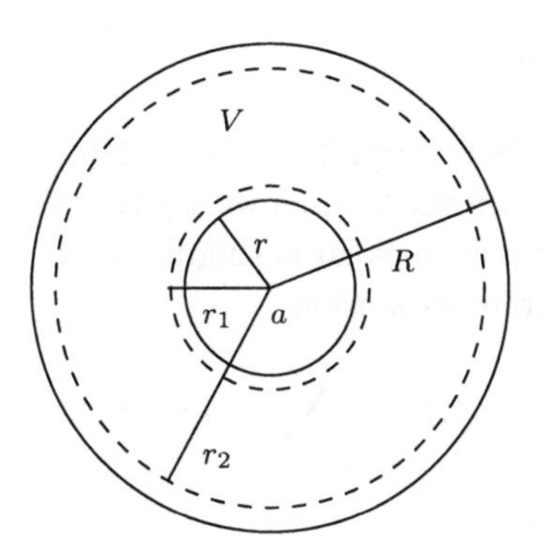

Fig. 3

and the series at the right end of the equation converges uniformly on γ_1.

If $\zeta \in \gamma_2$, then $|(z-a)/(\zeta-a)| < 1$. So we have

$$\frac{1}{\zeta - z} = \frac{1}{(\zeta - a)\left(1 - \dfrac{z-a}{\zeta - a}\right)} = \sum_{n=0}^{\infty} \frac{(z-a)^n}{(\zeta-a)^{n+1}},$$

and the series at the right end of the equation converges uniformly on γ_2. Substitute these two equations into (3.4) yields (3.2) and (3.3).

Now, we prove the uniqueness.

Suppose that $f(z)$ has another Laurent series expansion

$$f(z) = \sum_{n=-\infty}^{\infty} c_n'(z-a)^n, \quad (r < |z-a| < R) \tag{3.5}$$

on U, and it converges uniformly to $f(z)$ on $|z-a| = \rho\,(r < \rho < R)$. Multiply $1/(z-a)^{m+1}$ on both sides of (3.5) and integrate on $|z-a| = \rho$. Since the series converges uniformly and

$$\int_{|z-a|=\rho} (z-a)^k\, dz = \begin{cases} 2\pi i, & \text{if } k = -1, \\ 0, & \text{if } k \neq -1, \end{cases}$$

we have

$$\int_{|z-a|=\rho} \frac{f(z)dz}{(z-a)^{m+1}} = \sum_{n=-\infty}^{\infty} c_n' \int_{|z-a|=\rho} (z-a)^{n-m-1}\, dz = 2\pi i c_m'.$$

Thus, $c_m' = c_m\,(m = 0, \pm 1, \pm 2, \cdots)$ and this proves the uniqueness.

3.2 Isolated Singularity

If a function $f(z)$ is holomorphic on $U = D(a, R) \setminus \{a\}$, an open disc obtained by removing a from D, then we say that the point a is an isolated singularity of $f(z)$. If a is an isolated singularity of $f(z)$, then by Theorem 3.2, $f(z)$ has the Laurent expansion

$$f(z) = \sum_{n=-\infty}^{\infty} c_n (z-a)^n,$$

on $0 < |z - a| < R$, where

$$c_n = \frac{1}{2\pi i} \int_{|\zeta - a| = \rho} \frac{f(\zeta) d\zeta}{(\zeta - a)^{n+1}}, \quad (0 < \rho < R),$$

$$n = 0, \pm 1, \pm 2, \cdots.$$

By the discussion of the last section, we have $f(z) = \varphi(z) + \psi(z)$, where

$$\varphi(z) = \sum_{n=0}^{\infty} c_n (z-a)^n \tag{3.6}$$

is holomorphic on $|z - a| < R$. The series (3.6) is the holomorphic part of the Laurent expansion of $f(z)$ at the point a. The series

$$\psi(z) = \sum_{n=1}^{\infty} c_{-n} (z-a)^{-n} \tag{3.7}$$

is holomorphic on $0 < |z - a| < \infty$. The series (3.7) is the principle part of the Laurent expansion of $f(z)$ at the point a.

There are three possibilities for $\lim_{z \to a} f(z)$:

(1) If $\lim_{z \to a} f(z)$ exists and is finite, then by the Riemann Theorem (Theorem 2.11 in Section 2.3), $f(z)$ has an analytic continuation to $D(a, R)$. (This can be proved directly by using Theorem 3.2. We leave the proof to the reader as an exercise.) Thus, every c_{-n} is zero in (3.7). Conversely, if every c_{-n} in (3.7) is zero, then $f(z) = \varphi(z)$. It follows that $\lim_{z \to a} f(z) = \varphi(a)$. Therefore, $\lim_{z \to a} f(z)$ exists and is finite if and only if c_{-n} is zero for all n. In this case, the point a is called a *removable singularity*.

(2) The function $f(z)$ has an infinite limit at a if and only if there are only finite number of non-zero c_{-n}s. That is

$$\psi(z) = \frac{c_{-1}}{z-a} + \cdots + \frac{c_{-m}}{(z-a)^m}, \quad c_{-m} \neq 0.$$

So we have

$$\begin{aligned} f(z) &= \varphi(z) + \psi(z) \\ &= \frac{c_{-m}}{(z-a)^m} + \cdots + \frac{c_{-1}}{z-a} + c_0 + c_1(z-a) + \cdots \\ &= \frac{g(z)}{(z-a)^m}, \end{aligned}$$

where

$$g(z) = c_{-m} + c_{-m+1}(z-a) + \cdots, \quad c_{-m} = g(a) \neq 0.$$

The point a in this case is said to be a *pole* with order m. It is called a *simple pole* if $m = 1$.

The sufficiency is obvious in part (2), we now prove the necessity.

Since $\lim_{z \to a} f(z) = \infty$, there exists a $\delta > 0$ such that $f(z) \neq 0$ in $0 < |z-a| < \delta$. Thus, the function $F(z) = 1/f(z)$ is holomorphic and non-zero in $0 < |z-a| < \delta$. Moreover, $\lim_{z \to a} F(z) = 0$. By (1), the point a is a removable singularity of $F(z)$ and is also a zero of $F(z)$.

Suppose that a is a zero of order m. Then $F(z) = (z-a)^m \lambda(z)$, where $\lambda(z)$ is holomorphic on $|z-a| < \delta$. Choose a neighborhood of a such that $\lambda(z)$ does not vanish in it. Without loss of generosity, we can let this neighborhood to be $|z-a| < \delta$. Then, $1/\lambda(z)$ is holomorphic and does not vanish on this neighborhood. The Taylor expansion of $1/\lambda(z)$ on $|z-a| < \delta$ is

$$\frac{1}{\lambda(z)} = c_{-m} + c_{-m+1}(z-a) + \cdots, \quad c_{-m} \neq 0.$$

Therefore,

$$\begin{aligned} f(z) &= \frac{1}{F(z)} = \frac{1}{(z-a)^m \lambda(z)} \\ &= \frac{c_{-m}}{(z-a)^m} + \cdots + \frac{c_{-1}}{z-a} + c_0 + c_1(z-a) + \cdots. \end{aligned}$$

By the uniqueness of the Laurent expansion, (2) is proved.

From (1) and (2), we have

(3) The necessary and sufficient condition for $\lim_{z \to \infty} f(z)$ to not exist is that there are infinitely many non-zero c_{-n}s in (3.7). The point a in this case is a *essential singularity*. For example, $z = 0$ is an essential singularity of function $f(z) = e^{1/z}$. Indeed, since $\lim_{z=x \to 0^+} e^{1/z} = +\infty$ and $\lim_{z=x \to 0^-} e^{1/z} = 0$, it follows that $\lim_{z \to 0} f(z)$ does not exists.

We prove the following important theorem about essential singularity.

Theorem 3.3 ***(Weierstrass Theorem)*** *Let a be an essential singularity of $f(z)$. If $\delta > 0$ is given, then for any finite complex number A and any $\varepsilon > 0$, there exists a point z in $0 < |z-a| < \delta$ such that $|f(z) - A| < \varepsilon$. In other words, the values of $f(z)$ in the neighborhood of an essential singularity are dense in $\mathbb{C}$.*

Proof Suppose the contrary. Then there exist a finite complex number A and a positive number ε such that $|f(z) - A| > \varepsilon$ in $0 < |z-a| < \delta$. Thus

$$F(z) = \frac{f(z) - A}{z - a}$$

is holomorphic on $0 < |z-a| < \delta$ and $F(z) \to \infty$ as $z \to a$. Therefore a is a pole of $f(z)$. By (2),

$$F(z) = \frac{c_{-m}}{(z-a)^m} + \cdots + \frac{c_{-1}}{z-a} + c_0 + c_1(z-a) + \cdots ,$$

Hence

$$\begin{aligned} f(z) = &\frac{c_{-m}}{(z-a)^{m-1}} + \cdots + \frac{c_{-2}}{z-a} \\ &+ (A + c_{-1}) + c_0(z-a) + \cdots . \end{aligned}$$

It follows that the point a is either a pole of $f(z)$ with order m (when $m > 1$), or a removable singularity of $f(z)$ (when $m = 1$). This contradicts the assumption of a and we complete the proof.

Weierstrass Theorem characterized the distribution of the values of $f(z)$ near an essential singularity. In 1879, Picard proved a more generalized and profound theorem which is referred to as the *Great Picard Theorem*: An analytic function f assumes each finite complex value, with the exception of possibly one, an infinite number of times in each neighborhood of its essential singularity a. The proof of this theorem will be given in Chapter 5.

If $f(z)$ is a holomorphic function on the annulus $V = \{z | \, R < |z| < \infty, R > 0\}$, then $z = \infty$ is a isolated singularity of $f(z)$. The function $\zeta = 1/z$ maps the neighborhoods of $z = \infty$ to the neighborhoods of $\zeta = 0$. Thus $g(\zeta) = f(z) = f(1/\zeta)$ is holomorphic on $0 < |\zeta| < 1/R$ and has the

Laurent expansion

$$g(\zeta) = \sum_{n=-\infty}^{\infty} c_{-n}\zeta^n = \sum_{n=0}^{\infty} c_{-n}\zeta^n + \sum_{n=1}^{\infty} c_n\zeta^{-n} = \varphi(\zeta) + \psi(\zeta),$$

where $\varphi(\zeta)$ is the holomorphic part and $\psi(\zeta)$ is the principle part. Therefore

$$f(z) = \sum_{n=-\infty}^{\infty} \frac{c_{-n}}{z^n} = \sum_{n=0}^{\infty} \frac{c_{-n}}{z^n} + \sum_{n=1}^{\infty} c_n z^n = \varphi_0(z) + \psi_0(z),$$

where $\varphi_0(z)$ is the holomorphic part and $\psi_0(z)$ is the principle part. Hence, we have

(1) if $z = \infty$ is a removable singularity of $f(z)$, then

$$f(z) = c_0 + \frac{c_{-1}}{z} + \frac{c_{-2}}{z^2} + \cdots;$$

(2) if $z = \infty$ is a pole of $f(z)$ with degree m, then

$$f(z) = \sum_{n=0}^{\infty} \frac{c_{-n}}{z^n} + c_1 z + c_2 z^2 + \cdots + c_m z^m, \quad c_m \neq 0;$$

(3) if $z = \infty$ is an essential singularity of $f(z)$, then

$$f(z) = \sum_{n=0}^{\infty} \frac{c_{-n}}{z^n} + \sum_{n=1}^{\infty} c_n z^n.$$

3.3 Entire Functions and Meromorphic Functions

If a function $f(z)$ is holomorphic on $\mathbb{C}$ except the point $z = \infty$, then $f(z)$ is called an *entire function*. The function $f(z)$ has the Taylor expansion

$$f(z) = \sum_{n=0}^{\infty} c_n z^n \tag{3.8}$$

on $\mathbb{C}$. Since $z = \infty$ is an isolated singularity, by the uniqueness of Laurent expansion, (3.8) is also the Laurent expansion of $f(z)$ on the neighborhood of $z = \infty$. Thus, there are three possibilities:

(1) If $z = \infty$ is a removable singularity of $f(z)$, then by the Liouville Theorem (Theorem 2.10 in Section 2.3), $f(z)$ is constant.

(2) If $z = \infty$ is a pole of $f(z)$ with order m, then $c_n = 0$ for $n > m$. In this case, $f(z)$ is a polynomial with degree m,

$$f(z) = c_0 + c_1 z + c_2 z^2 + \cdots + c_m z^m, \quad c_m \neq 0.$$

(3) If $z = \infty$ is an essential singularity of $f(z)$, then

$$f(z) = c_0 + c_1 z + c_2 z^2 + \cdots + c_n z^n + \cdots,$$

and there are infinitely many $c_n \neq 0$, where $n \geq 1$. Functions of this type are called *transcendental entire functions*. Examples include e^z, $\sin z$, $\cos z$, etc.

If a function $f(z)$ only has poles on $\mathbb{C}$ besides the infinity point (the number of poles can be either finite or infinity,) then $f(z)$ is called a *meromorphic function*. An entire function is a meromorphic function. Rational functions $f(z) = P_n(z)/Q_m(z)$, where $P_n(z)$ and $Q_m(z)$ are reduced polynomials, are also meromorphic functions. Let

$$P_n(z) = a_0 + a_1 z + \cdots + a_n z^n, \quad a_n \neq 0,$$

$$Q_m(z) = b_0 + b_1 z + \cdots + b_m z^m, \quad b_m \neq 0.$$

The zeros of $Q_m(z)$ are the poles of $f(z)$. Since

$$f(z) = \frac{1}{z^{m-n}} \frac{a_n + \dfrac{a_{n-1}}{z} + \cdots + \dfrac{a_0}{z^n}}{b_m + \dfrac{b_{m-1}}{z} + \cdots + \dfrac{b_0}{z^m}},$$

we have

$$\lim_{z \to \infty} f(z) = \begin{cases} \dfrac{a_n}{b_m}, & \text{if } m = n, \\ \infty, & \text{if } n > m, \\ 0, & \text{if } n < m. \end{cases}$$

It follows that, $z = \infty$ is either a removable singularity or a pole of $f(z)$.

On the other hand, we have

Theorem 3.4 *If $z = \infty$ is a removable singularity or a pole of a holomorphic function $f(z)$, then $f(z)$ must be a rational function.*

Proof Since $z = \infty$ is a removable singularity or a pole of $f(z)$, there exists a $R > 0$ such that $f(z)$ is holomorphic on $R < |z| < \infty$. Let $p(z)$ be the principle part of the Laurent expansion of $f(z)$ in a neighborhood of $z = \infty$. If $z = \infty$ is a removable singularity of $f(z)$, then $p(z)$ equal to zero identically. If $z = \infty$ is a pole of $f(z)$, then $p(z)$ is a polynomial.

On the disc $|z| \leq R$, $f(z)$ can only have finitely many poles. If not, then by the Bolzano-Weierstrass Theorem, there is a limit point z_0 of the infinitely many poles of $f(z)$. Since z_0 in $|z| \leq R$, z_0 is a non-isolated singularity. This is impossible because $f(z)$ is a meromorphic function. Suppose $z_1, z_2, \cdots, z_k$ are poles of $f(z)$. In a neighborhood of each z_i, the principle part of the Laurent expansion of $f(z)$ at z_i is

$$\psi_i(z) = \frac{c_{-1}^{(i)}}{z - z_i} + \cdots + \frac{c_{-m}^{(i)}}{(z - z_i)^m}, \quad i = 1, 2, \cdots, k.$$

The holomorphic part of the Laurent expansion of $f(z)$ at z_i is denoted by $\varphi_i(z)$, $(i = 1, 2, \cdots, k)$. So the function

$$F(z) = f(z) - p(z) - \sum_{i=1}^{k} \psi_i(z)$$

is holomorphic on $\mathbb{C}$ except the points $z_1, z_2, \cdots, z_k$ and ∞. These points are removable singularities of $F(z)$. Indeed, Since $\lim_{z \to z_i}(f(z) - \psi_i(z)) = \varphi_i(z)$ and $\sum_{m \neq i} \psi_m(z) - p(z)$ is holomorphic at z_i, the limit $\lim_{z \to z_i} F(z)$ exists and is finite. The function $f(z) - p(z)$ is the holomorphic part of Laurent expansion of $f(z)$ at the neighborhood of $z = \infty$, so $\lim_{z \to \infty}(f(z) - p(z))$ exists and is finite and $\lim_{z \to \infty} \sum_{i=1}^{k} \psi_i(z) = 0$. Therefore, $\lim_{z \to \infty} F(z)$ is finite and $F(z)$ is holomorphic on $\mathbb{C}$. By Liouville Theorem, $F(z)$ is a constant c. Hence,

$$f(z) = c + p(z) + \sum_{i=1}^{k} \psi_i(z)$$

is a rational function and the theorem follows.

A meromorphic function is called a *transcendental meromorphic function* if it is not a rational function. For transcendental meromorphic functions, the point $z = \infty$ is either an essential singularity or a limit point of poles.

The group of holomorphic automorphisms on the unit disc was given in Section 2.5 (Theorem 2.20). Now we observe the group of holomorphic automorphisms on the complex plane $\mathbb{C}$ and the extended complex plane $\mathbb{C}^*$ (equivalently, the Riemann sphere S^2).

First we find the group of holomorphic automorphisms on $\mathbb{C}$, $\mathrm{Aut}(\mathbb{C})$.

Let $\alpha(z) \in \mathrm{Aut}(\mathbb{C})$. Then $\alpha(z)$ maps ∞ to ∞. Since the map is an automorphism, it is one-to-one. Thus, $z = \infty$ is a simple pole of $\alpha(z)$. According to the preceding result, $\alpha(z)$ must be a polynomial of degree one, $\alpha(z) = az + b$, where $a, b \in \mathbb{C}$ and $a \neq 0$. Conversely, it is easy to see that $az + b \in \mathrm{Aut}(\mathbb{C})$ if $a, b \in \mathbb{C}$ and $a \neq 0$. So $\mathrm{Aut}(\mathbb{C})$ consists of all linear transformations $az+b$ with $a, b \in \mathbb{C}$ and $a \neq 0$. In other words, every element of Aut($\mathbb{C}$) is the composition of a translation $\alpha(z) = z + b$ and a dilation $\alpha(z) = az$.

Next, we find the group of holomorphic automorphism on $\mathbb{C}^*$, Aut($\mathbb{C}^*$).

Let $\alpha(z) \in$ Aut($\mathbb{C}^*$) and $\alpha(\infty) = \infty$. Since $\alpha(z)$ is an automorphism, it is one-to-one. So $\alpha(z) \in$ Aut($\mathbb{C}$) on $\mathbb{C}$. Hence, $\alpha(z) = cz + d$ where $c, d \in \mathbb{C}$ and $c \neq 0$. It is easy to verify that , if $a, b, c, d \in \mathbb{C}$ and $ad - bc \neq 0$, then $\alpha(z) = (az + b)/(cz + d) \in$ Aut($\mathbb{C}^*$).

If $\alpha(z) \in$ Aut($\mathbb{C}^*$) and $\alpha(\infty) \neq \infty$, then $\beta(z) = 1/(\alpha(z) - \alpha(\infty)) \in$ Aut($\mathbb{C}^*$) and $\beta(\infty) = \infty$. Thus $\beta(z) = cz + d$, where $c, d \in \mathbb{C}$ and $c \neq 0$. Therefore, $cz + d = 1/(\alpha(z) - \alpha(\infty))$. Solve for $\alpha(z)$ from this equation, we get that $\alpha(z) = (az + b)/(cz + d)$ where $a = \alpha(\infty)c$, $b = d\alpha(\infty) + 1$. Hence, Aut($\mathbb{C}^*$) consists of all fractional linear transformations $(az + b)/(cz + d)$ where $ad - bc = 1$. In other words, the elements of Aut($\mathbb{C}^*$) are compositions of a translation $\alpha(z) = z + b$, a dilation $\alpha(z) = az$ and an inversion $\alpha(z) = 1/z$. Under the one-to-one correspondence between $(az + b)/(cz + d)$ and 2×2 matrix

$$\begin{pmatrix} a & b \\ c & d \end{pmatrix}$$

Aut($\mathbb{C}^*$) is isomorphic to the group

$$\left\{ \begin{pmatrix} a & b \\ c & d \end{pmatrix} \middle| \det \begin{pmatrix} a & b \\ c & d \end{pmatrix} = 1 \right\} / \{\pm I\}$$

where

$$I = \begin{pmatrix} 1 & 0 \\ 0 & 1 \end{pmatrix}$$

is the identity matrix. This group is called the *group of Möbius transformations*. In fact, it is $SL(2, \mathbb{C})/\{\pm I\}$, where $SL(2, \mathbb{C})$ is the special linear group of order two.

The following theorem is very important in complex analysis.

Theorem 3.5 ***(Uniformization Theorem or Poincaré-Koebe Theorem)*** *Any simply connected Riemann surface is one-to-one holomorphically equivalent to one of these three regions: the unit disc, the complex plane and the extended complex plane (Riemann sphere S^2).*

We will talk more about Riemann surfaces in the next chapter. The automorphism groups of the unit disc, $\mathbb{C}$ and $\mathbb{C}^*$ are determined by the Theorem

2.20 in Section 2.5 and the result discussed above. It follows that, the automorphism groups of the regions which are holomorphically equivalent in a one-to-one manner to simply connected Riemann surfaces are also determined in the same way. In the next chapter, we will prove the important *Riemann Mapping Theorem*:

Theorem 3.6 *Any simply connected regions whose boundary contain at least two points, are one-to-one holomorphically equivalent to the unit disc.*

In other words, under the one-to-one holomorphic equivalence, the unit disc, $\mathbb{C}$ and $\mathbb{C}^*$ are the only three simply connected regions. The geometry of these three regions will be discussed in Chapter 5.

We can say that, Poincaré-Koebe Uniformization Theorem is one of the most important and beautiful theorems in complex analysis.

3.4 Weierstrass Factorization Theorem, Mittag-Leffler Theorem and Interpolation Theorem

In this section, we will prove three constructive theorems.

Since entire functions are holomorphic except at the infinity, they can be expressed by the Taylor series (3.8) in Section 2.3. According to the discussion in Section 3.3, if $z = \infty$ is a pole of an entire function $f(z)$, then $f(z)$ is a polynomial. So entire functions can be considered as a natural generalization of polynomials. They are the polynomials of the form (3.8) with infinite degree. An explicit way to express a polynomial is by its roots (zeros). If $a_1, \cdots, a_n$ are the roots of a n-th degree polynomial $P_n(z)$, then $P_n(z)$ can be written as $A(z-a_1)\cdots(z-a_n)$, where A is a complex constant. This expression is called a *factorization* of $P_n(z)$. Can transcendental entire functions also be factorized? In other words, if $a_1, a_2, \cdots, a_n, \cdots$ are the infinite zeros of an entire function, can this function be expressed as

$$A(z-a_1)\cdots(z-a_n)\cdots = A\prod_{k=1}^{\infty}(z-a_i)\,?$$

The question cannot be answered directly. Since the product is infinity, whether it is convergent has to be considered. The answer for this question is the Weierstrass Factorization Theorem. Before we state this theorem, let's have a brief discussion about infinite products.

For a complex sequence $\{u_n\}$, consider the product

$$p_n = \prod_{k=1}^{n}(1+u_k).$$

If $1+u_k \neq 0, (k=1,2,\cdots)$, $\lim_{n\to\infty} p_n = p \neq 0$ and p is finite, then we say that the infinite product

$$\prod_{n=1}^{\infty}(1+u_n) \tag{3.9}$$

is *convergent* and it converges to p, denoted $p=\prod_{n=1}^{\infty}(1+u_n)$. Otherwise, we say that (3.9) is *divergent.*

Since $1+x \leq e^x$ for $x \geq 0$, we have

$$\begin{aligned}|u_1|+|u_2|+\cdots+|u_n| &\leq (1+|u_1|)(1+|u_2|)\cdots(1+|u_n|)\\ &\leq e^{|u_1|+|u_2|+\cdots+|u_n|}.\end{aligned}$$

Hence, the convergence of $\sum_{n=1}^{\infty}|u_n|$ implies the convergence of $\prod_{n=1}^{\infty}(1+|u_n|)$ and vice versa. If $\sum_{n=1}^{\infty}|u_n|$ converges, then we say that (3.9) is *absolutely convergent.* Therefore, we have the following assertion:

An absolutely convergent infinite product must be convergent, and its value is independent of the changing of the order of its factors.

We omit the proof here.

Next, we discuss the factorization of entire functions.

If an entire function $f(z)$ has no zero, then $f(z)=e^{\varphi(z)}$, where $\varphi(z)$ is an entire function. Indeed, by the fact that $f'(z)/f(z)$ is holomorphic on the whole plane, it is the derivative of an entire function $\varphi(z)$. We can find that the derivative of $f(z)e^{-\varphi(z)}$ is zero. It follows that $f(z)$ is a constant multiple of $e^{\varphi(z)}$ and this constant can be absorbed into $\varphi(z)$.

Suppose an entire function $f(z)$ only has finitely many zeros 0, a_1, a_2, $\cdots$, a_n, where $a_i \neq 0$ $(i=1,\cdots,n)$ with orders m, m_1, m_2, $\cdots$, m_n respectively. Let

$$p(z)=z^m\left(1-\frac{z}{z_1}\right)^{m_1}\cdots\left(1-\frac{z}{a_n}\right)^{m_n}.$$

Then $h(z)=f(z)/p(z)$ has $z=0, a_i\,(i=1,\cdots,n)$ as its removable singularities. So $h(z)$ is an entire function without zero. Hence $h(z)=e^{\psi(z)}$

where $\psi(z)$ is an entire function. Therefore,

$$f(z) = z^m \left(1 - \frac{z}{a_1}\right)^{m_1} \cdots \left(1 - \frac{z}{a_n}\right)^{m_n} e^{\psi(z)},$$

In words, $f(z)$ can be expressed as the product of a polynomial and an entire function without zero. The zeros of the polynomial are also the zeros of $f(z)$ and they have the same orders.

Suppose an entire function $f(z)$ has infinitely many zeros and is not equal to zero identically. Since the zeros of $f(z)$ are countable, they can be arranged into a sequence

$$a_1, a_2, \cdots, a_n, \cdots$$

by the values of the modulus of each term in an increasing manner,

$$0 < |a_n| \leq |a_{n+1}|, \quad \lim_{n\to\infty} |a_n| = \infty. \tag{3.10}$$

We exclude $z = 0$ here, it will be treated differently.

Since $\lim_{n\to\infty} |a_n| = \infty$, for any positive number R, there exists a sequence of positive integers $k_1, k_2, \cdots, k_n, \cdots$ such that $\sum_{n=1}^{\infty}(R/|a_n|)^{k_n+1}$ is convergent. Such a sequence does exist. For instance, let $k_n = n - 1$. Consider the infinite product

$$\prod_{n=2}^{\infty} \left(1 - \frac{z}{a_n}\right) \exp\left\{\frac{z}{a_n} + \frac{1}{2}\left(\frac{z}{a_n}\right)^2 + \cdots + \frac{1}{k_n}\left(\frac{z}{a_n}\right)^{k_n}\right\}. \tag{3.11}$$

Let

$$P_n(z) = \frac{z}{a_n} + \frac{1}{2}\left(\frac{z}{a_n}\right)^2 + \cdots + \frac{1}{k_n}\left(\frac{z}{a_n}\right)^{k_n},$$

$$Q_n(z) = \log\left(1 - \frac{z}{a_n}\right) + P_n(z),$$

$$E_n(z) = \left(1 - \frac{z}{a_n}\right) e^{P_n(z)} = e^{Q_n(z)}.$$

Then (3.11) becomes $\prod_{n=2}^{\infty} E_n(z)$.

For any fixed positive number R, choose a positive integer N, such that $|a_n| > 2R$ when $n \geq N$. Consider the infinite product $\prod_{n=N}^{\infty} E_n(z)$, since

$|z/a_n| \le 1/2$ for $|z| \le R$ and $n \ge N$, we have

$$\begin{aligned}|Q_n(z)| &\le \frac{1}{k_n+1}\left(\frac{|z|}{|a_n|}\right)^{k_n+1} + \frac{1}{k_n+2}\left(\frac{|z|}{|a_n|}\right)^{k_n+2} + \cdots \\ &\le \left(\frac{|z|}{|a_n|}\right)^{k_n+1} \frac{1}{1-\dfrac{|z|}{|a_n|}} \le 2\left(\frac{R}{|a_n|}\right)^{k_n+1}.\end{aligned}$$

Since $\sum_{n=1}^{\infty}(R/|a_n|)^{k_n+1}$ converges, the sequence $\sum_{n=N}^{\infty} Q_n(z)$ converge absolutely and uniformly on $|z| \le R$. Hence

$$\prod_{n=N}^{\infty} E_n(z) = \exp\left(\sum_{n=N}^{\infty} Q_n(z)\right)$$

converges uniformly on $|z| \le R$. By the Weierstrass Theorem (Theorem 3.1), this infinite product represents a non-zero holomorphic function on $|z| < R$. The zeros of

$$\left(1-\frac{z}{a_1}\right)\prod_{n=2}^{N-1} E_n(z)$$

are $a_1, \cdots, a_{N-1}$ and they are all inside $|z| \le 2R$. Hence, those a_n $(n = 1, 2, \cdots)$ in $|z| \le R$ are the zeros of

$$\left(1-\frac{z}{a_1}\right)\prod_{n=2}^{N-1} E_n(z)$$

and they are the only zeros in $|z| < R$.

For any z in $|z| < R$ and any n sufficiently large, we have $|Q_n(z)| < 1$. It is easy to prove that $|e^z - 1| \le (7/4)|z|$ when $|z| < 1$. Thus we have

$$\begin{aligned}|E_n(z) - 1| &= |e^{Q_n(z)} - 1| \\ &\le \frac{7}{4}|Q_n(z)| \le \frac{7}{2}\left(\frac{R}{|a_n|}\right)^n.\end{aligned}$$

Therefore, $\prod_{n=N}^{\infty} E_n(z)$ converge absolutely on $|z| \le R$ when N sufficiently large.

Hence, for the given complex sequence (3.10), there exists an entire function

$$g(z) = \left(1-\frac{z}{a_1}\right)\prod_{n=2}^{\infty} E_n(z)$$

with zeros a_n $(n = 1, 2, \cdots)$ and it converges absolutely. Moreover,

$$P_N(z) = \left(1 - \frac{z}{a_1}\right) \prod_{n=2}^{N} E_n(z)$$

converges uniformly to $g(z)$ on any disc $|z| < R$. Therefore, we have

Theorem 3.7 ***(Weierstrass Factorization Theorem)*** *Suppose $f(z)$ is an entire function, $z = 0$ is a zero of $f(z)$ with multiplicity m (m can be zero) and the rest of the zeros are $a_1, a_2, \cdots$ which satisfy $0 < |a_n| \le |a_{n+1}|$ and $\lim_{n\to\infty} |a_n| = \infty$. If for any $R > 0$, there exists a sequence of positive integers $k_1, k_2, \cdots, k_n, \cdots$ such that $\sum_{n=1}^{\infty} (R/|a_n|)^{k_n+1}$ is convergent, then*

$$f(z) = z^m e^{h(z)} \prod_{n=1}^{\infty} \left(1 - \frac{z}{a_n}\right) \cdot \exp\left\{\frac{z}{a_n} + \frac{1}{2}\left(\frac{z}{a_n}\right)^2 + \cdots + \frac{1}{k_n}\left(\frac{z}{a_n}\right)^{k_n}\right\}, \tag{3.12}$$

where $h(z)$ is an entire function. If $k_n = n - 1$, then

$$f(z) = z^m e^{h(z)} \prod_{n=1}^{\infty} \left(1 - \frac{z}{a_n}\right) \exp\left\{\frac{z}{a_n} + \frac{1}{2}\left(\frac{z}{a_n}\right)^2 + \cdots + \frac{1}{n-1}\left(\frac{z}{a_n}\right)^{n-1}\right\}.$$

Proof Construct an entire function $g(z)$ with zeros a_n $(n = 1, 2, \cdots)$. Then $z^m g(z)$ and $f(z)$ have the same zeros with the same multiplicities. Thus $0, a_i$ $(i = 1, 2, \cdots)$ are removable singularities of $H(z) = f(z)/(z^m g(z))$ and $H(z)$ has no zeros. Therefore, $H(z) = e^{h(z)}$, where $h(z)$ is an entire function. This proves the theorem.

Since $g(z)$ is not unique, the expression of $f(z)$ is also not unique.

We have the following assertion about the expressions of entire functions from the proof of Weierstrass Factorization Theorem:

Any meromorphic function can be represented as the ratio of two entire functions.

To prove this assertion, if $f(z)$ is a meromorphic function, then there exists an entire function $f_1(z)$ which has the poles of $f(z)$ as its zeros. Let $f_2(z) = f(z)f_1(z)$ and a be a pole of $f(z)$. Define $f_2(a) = \lim_{z\to a} f_2(z)$.

Then $f_2(z)$ is also an entire function. Therefore, $f(z) = f_2(z)/f_1(z)$ and the assertion follows.

Since any meromorphic function can be represented by the ratio of two entire functions and every entire function can be represented by an expression in the form of (3.12) according to the Weierstrass Factorization Theorem, it follows that every meromorphic function can be represented as the ratio of two expressions of the form (3.12). Hence, meromorphic functions can be represented explicitly by their zeros and poles.

Moreover, from the last section, if a meromorphic function $f(z)$ only has finitely many poles $a_1, a_2, \cdots, a_n$ and $z = \infty$ is its pole or removable singularity, then $f(z)$ is a rational function and

$$f(z) = c + p(z) + \sum_{j=1}^{n} \psi_j(z),$$

where c is a constant, $p(z)$ is a polynomial and $\psi_j(z)$ is the principle part of the Laurent series of $f(z)$ at the pole $z = a_j$ (Theorem 3.6).

For a transcendental meromorphic function $f(z)$, $z = \infty$ is either an essential singularity or a limit point of poles. If $z = \infty$ is an essential singularity, then $f(z)$ still has finitely many poles. Thus, $U(z) = f(z) - \sum_{j=1}^{n} \psi_j(z)$ is a transcendental entire function. Hence

$$f(z) = U(z) + \sum_{j=1}^{n} \psi_j(z).$$

If $z = \infty$ is a limit point of poles $a_1, a_2, \cdots, a_n, \cdots$ of $f(z)$, where $|a_n| \leq |a_{n+1}|$ and $\lim_{n\to\infty} |a_n| = \infty$, the principle part of Laurent expansion of $f(z)$ at a_j, $\psi_j(z)$, was given, then the next theorem answers the question of whether there is a meromorphic function $f(z)$ such that it has $a_1, a_2, \cdots, a_n, \cdots$ as its poles and has $\psi_1, \psi_2, \cdots, \psi_n, \cdots$ as the principle parts of its Laurent expansions at these poles respectively.

Theorem 3.8 ***(Mittag-Leffler Theorem)*** *There exists such a meromorphic function $f(z)$ whose poles are $a_1, a_2, \cdots, a_n, \cdots$ where $|a_n| \leq |a_{n+1}|$ and $\lim_{n\to\infty} |a_n| = \infty$, and the principle parts of the Laurent series at a_i $(1 \leq i < \infty)$ are $\psi_1(z), \psi_2(z), \cdots, \psi_n(z), \cdots$.*

Proof Choose a neighborhood U_i of a_i $(i = 1, 2, \cdots)$ such that $U_i \bigcap U_j = \emptyset$ if $i \neq j$. Let φ_i be a C^∞ function such that $\varphi_i = 1$ on a small neighborhood V_i of a_i, where $V_i \subset U_i$, and $\varphi_i = 0$ on the complement of U_i $(i = 1, 2, \cdots)$. Define $u = \sum_{i=1}^{\infty} \varphi_i \phi_i$ on $\mathbb{C} \setminus \{a_i\}_1^\infty$. Then u is a C^∞

function on $\mathbb{C} \setminus \{a_i\}_1^\infty$ and $u = \psi_i$ on $V_i \setminus \{a_i\}$. That is, u has the principle part as required near a_i, but u is not meromorphic. Let

$$A = \begin{cases} \dfrac{\partial u}{\partial \bar{z}}, & \text{if } z \in \mathbb{C} \setminus \{a_i\}_1^\infty, \\ 0, & \text{if } z = a_i, i = 1, 2, \cdots. \end{cases}$$

Since $u = \psi_i$ on $V_i \setminus \{a_i\}$, we have $\partial u/\partial \bar{z} = 0$ for $z \neq a_i$. That is, $A = 0$ on $V_i \setminus \{a_i\}$. By the definition of A, $A = 0$ if $z = a_i$. Therefore, A is a continuous function. Similarly, we can prove that A is also a C^∞ function. Hence, by the solution of one dimensional $\bar{\partial}$-problem (Theorem 2.4 in Section 2.1), the $\bar{\partial}$-problem

$$\frac{\partial v}{\partial \bar{z}} = A$$

has a C^∞ solution v and it can be expressed as (2.4) in Section 2.1. Thus, $f = u - v$ is the meromorphic function we wanted. Obviously,

$$\frac{\partial f}{\partial \bar{z}} = \frac{\partial(u - v)}{\partial \bar{z}} = 0$$

when $z \in \mathbb{C} \setminus \{a_i\}_1^\infty$. Therefore, f is holomorphic when $z \in \mathbb{C} \setminus \{a_i\}_1^\infty$. Since $v \in C^\infty(\mathbb{C})$, by the definition of u, f has the principle part $\psi_i(z)$ for z near a_i. This proves the theorem.

The proof of Mittag-Leffler Theorem is simple and clean when the solution of $\bar{\partial}$-problem is used as a tool, but the meromorphic function can be expressed more explicitly if we use classical analysis to prove the theorem.

Theorem 3.9 ***(Mittag-Leffler Theorem)*** *Let $f(z)$ be a meromorphic function. Let $a_n\,(n = 1, 2, \cdots)$ be the poles of $f(z)$ where $|a_n| \leq |a_{n+1}|$ and $\lim_{n\to\infty} |a_n| = \infty$. Then $f(z)$ can be written as*

$$f(z) = U(z) + \sum_{n=1}^{\infty} \{\Psi_n(z) - P_n(z)\},$$

where $\Psi_n(z)$ is the principle part of $f(z)$ at the pole $z = a_n$, $P_n(z)\,(n = 1, 2, \cdots)$ are polynomials and $U(z)$ is an entire function.

Proof Choose $\varepsilon_n > 0\,(n = 1, 2, \cdots)$ such that $\sum_{n=1}^\infty \varepsilon_n$ is convergent. If $a_1 = 0$, then we can choose $P_1(z) = 0$. For $a_n\,(\neq 0)$, $\Psi_n(z)$ is a polynomial of $1/(z - a_n)$ and is holomorphic in $|z| < |a_n|$. Thus, it has Taylor expansion

$$\Psi_n(z) = \sum_{k=0}^{\infty} \frac{\Psi_n^{(k)}(0)}{k!} z^k,$$

in $|z| < |a_n|$. This series converges uniformly to $\Psi_n(z)$ in $|z| < (1/2)|a_n|$. So there exists λ_n such that

$$\left|\Psi_n(z) - \sum_{k=0}^{\lambda_n} \frac{\Psi_n^{(k)}(0)}{k!} z^k\right| < \varepsilon_n.$$

Let

$$P_n(z) = \sum_{k=0}^{\lambda_n} \frac{\Psi_n^{(k)}(0)}{k!} z^k$$

and R be an arbitrary positive number. Choose a positive integer $N = N(R)$ such that $|a_n| > 2R$ if $n > N$ and $|a_n| \leq 2R$ if $n \leq N$. Then we have

$$|\Psi_n(z) - P_n(z)| < \varepsilon_n$$

when $n > N$ and $|z| < R\,(|z| < (1/2)|a_n|)$.

Since $\sum_{n=1}^{\infty} \varepsilon_n$ converges, we have

$$\sum_{n=N+1}^{\infty} \{\Psi_n(z) - P_n(z)\}$$

converges uniformly on $|z| < R$. The pole of $\Psi_n(z)$, $z = a_n$, is not in $|z| < R$ when $n > N$. By the Weierstrass Theorem (Theorem 3.3),

$$\Phi_N(z) = \sum_{n=N+1}^{\infty} \{\Psi_n(z) - P_n(z)\}$$

is holomorphic on $|z| < R$. Thus,

$$\varphi(z) = \sum_{n=1}^{N} \{\Psi_n(z) - P_n(z)\} + \Phi_N(z)$$

has a_n, where $|a_n| < R\,(n = 1, 2, \cdots)$, as its poles, and $\Psi_n(z)\,(n = 1, 2, \cdots)$ as the corresponding principle parts. Since R is arbitrary, $\varphi(z)$ is a meromorphic function with poles $a_1, a_2, \cdots, a_n, \cdots$ and the corresponding principle parts $\Psi_1(z), \Psi_2(z), \cdots, \Psi_n(z), \cdots$. Define $U(z) = f(z) - \varphi(z)$ and $U(a_n) = \lim_{z \to a_n} \{f(z) - \varphi(z)\}$. Then, $U(z)$ is an entire function and the theorem follows.

For any given m points $z_1, \cdots, z_m$ and m complex numbers $a_1, \cdots, a_m$, we can find a polynomial $p(z)$ such that the value of $p(z)$ is a_j at $z_j\,(j = 1, 2, \cdots, m)$ by solving $p(z_j) = a_j\,(j = 1, \cdots, m)$ for the coefficients of $p(z)$. Similarly, for any given m points $z_1, \cdots, z_m$ and complex values $a_{j,k}\,(j =$

$1, \cdots, m$, $0 \le k \le n_j - 1)\, n_j \ge 1, n_j \in \mathbb{N}$, we can find a polynomial $p(z)$ such that $(p^{(k)}(z_j))/(k!) = a_{j,k}$. In other words, we can find a polynomial $p(z)$ such that the coefficients of the first n_j terms of the Taylor expansion of $p(z)$ at z_j are equal to the given numbers.

Now, we prove the general Interpolation Theorem.

Theorem 3.10 ***(Interpolation Theorem)*** *Let $\{z_1, z_2, \cdots\}$ be a discrete set in $\mathbb{C}$, $n_1, n_2, \cdots$ be a sequence of positive integers with each $n_i \ge 1, (i = 1, 2, \cdots)$ and $\{a_{j,k}\}\,(j \ge 1, 0 \le k \le n_j - 1)$ be a sequence of complex numbers. Then there exists an entire function $g(z)$ such that*

$$g^{(k)}(z_j) = k! a_{j,k} \quad (j \ge 1, 0 \le k \le n_j - 1).$$

In other words, if a sequence $\{z_j\}$ and the coefficients of the first n_j terms of the Taylor expansion at every z_j are given, then there exists an entire function that has such a Taylor expansion at these points.

Proof According to Theorem 3.7 (Weierstrass Factorization Theorem), we can find an entire function $f(z)$ such that $z = z_j$ is a zero of $f(z)$ with multiplicity n_j. Since $z_1, z_2, \cdots, z_n, \cdots$ is a discrete sequence, we can find a sequence $\varepsilon_1, \varepsilon_2, \cdots, \varepsilon_n, \cdots$ with each $\varepsilon_i > 0$ such that the discs with center z_j and radius $2\varepsilon_j$, $D(z_j, 2\varepsilon_j)\,(j = 1, 2, \cdots)$, are mutually exclusive. Suppose that

$$P_j(z) = \sum_{0 \le k \le n_j - 1} a_{j,k}(z - z_j)^k, \; j \ge 1.$$

For every j, let $\varphi_j \in C^\infty$ be such that $\operatorname{supp} \varphi_j \subset \bar{D}(z_j, 2\varepsilon_j)$ and $0 \le \varphi_j \le 1$, also, $\varphi_j \equiv 1$ on $\bar{D}(z_j, \varepsilon_j)$.

Let $\psi(z) \in C^\infty$. Construct the function

$$g(z) = \sum_{j \ge 1} P_j(z)\varphi_j(z) - f(z)\psi(z). \tag{3.13}$$

Since the supports of φ_j $(j = 1, 2, \cdots)$ are mutually exclusive on $\mathbb{C}$, for every $z \in \mathbb{C}$, $\sum_{j \ge 1} P_j(z)\varphi_j(z)$ has at most one non-zero term, so this sum is well defined. We want to find a ψ such that $g(z)$ is an entire function. In other words, $\partial g / \partial \bar{z} = 0$. If $z \in \mathbb{C}$, then this is equivalent to

$$\sum_{j \ge 1} P_j(z) \frac{\partial \varphi_j(z)}{\partial \bar{z}} = f(z) \frac{\partial \psi}{\partial \bar{z}}.$$

Let

$$h(z) = \sum_{j\geq 1} P_j(z)\frac{\partial \varphi_j(z)}{\partial \bar{z}}.$$

Then $h(z) \equiv 0$ on $\bigcup_j \overline{D(z_j, \varepsilon_j)}$. Choose $h(z)/f(z) = 0$ at $z = z_j$ then $h(z)/f(z)$ is C^∞ on $\mathbb{C}$. The function $h(z)$ has mutually exclusive compact supports. By Theorem 2.4 in Section 2.1, the $\bar{\partial}$-problem $\partial\psi/\partial\bar{z} = h/f$ has a C^∞ solution ψ. For such a ψ, (3.13) defined an entire function and $\partial\psi/\partial\bar{z} = 0$ on $\bigcup_j \overline{D(z_j, \varepsilon_j)}$. Thus ψ is holomorphic near z_j. Since the zero of $f(z)$ at $z = z_j$ has multiplicity n_j, from the definition of $g(z)$, (3.13), we have

$$g^{(k)}(z_j) = P_j^k(z_j) = k!a_{j,k}$$

when $0 \leq k \leq n_j - 1$. This proves the theorem.

For an arbitrary power series, we may not be able to find an entire function whose Taylor series is this given power series. But Theorem 3.10 tells us that, for an arbitrary polynomial and an arbitrary large number n (as long as n is finite), we can always find an entire function such that the given polynomial is the first n terms of its Taylor series. This is the crucial difference between polynomial and entire function, and Theorem 3.10 is a profound theorem.

3.5 Residue Theorem

Let $f(z)$ be a holomorphic function on $D(a,r) \setminus \{a\}$ and a be an isolated singularity of $f(z)$. The *residue* of $f(z)$ at a is defined by

$$\text{Res}(f,a) = \frac{1}{2\pi i}\int_{|z-a|=\rho} f(z)\,dz,$$

where $0 < \rho < r$. Since $f(z)$ has Laurent series

$$f(z) = \sum_{n=-\infty}^{\infty} c_n(z-a)^n$$

at $D(a,r) \setminus \{a\}$, we have $\text{Res}(f,a) = c_{-1}$.

If $z = \infty$ is an isolated singularity of $f(z)$ and $f(z)$ is holomorphic in

$R < |z| < \infty$, then we define the residue of $f(z)$ at $z = \infty$ as

$$\operatorname{Res}(f, \infty) = -\frac{1}{2\pi i} \int_{|z|=\rho} f(z)\, dz,$$

where $R < \rho < \infty$. Since $f(z)$ has Laurent expansion

$$f(z) = \sum_{n=-\infty}^{\infty} c_n z^n$$

on the neighborhood of $z = \infty$, we have $\operatorname{Res}(f, \infty) = c_{-1}$.

If $a(\neq \infty)$ is a pole of $f(z)$ with multiplicity m, then in a neighborhood of a, $f(z)$ can be written as

$$f(z) = \frac{1}{(z-a)^m} g(z),$$

where $g(z)$ is holomorphic at $z = a$ and $g(a) \neq 0$. Thus

$$g(z) = \sum_{n=0}^{\infty} \frac{1}{n!} g^{(n)}(a)(z-a)^n.$$

Hence

$$\operatorname{Res}(f, a) = c_{-1} = \frac{1}{(m-1)!} g^{(m-1)}(a).$$

Since

$$g^{(m-1)}(a) = \lim_{z \to a} \frac{d^{m-1}}{dz^{m-1}} \{(z-a)^m f(z)\},$$

we have

$$\operatorname{Res}(f, a) = \frac{1}{(m-1)!} \lim_{z \to a} \frac{d^{m-1}}{dz^{m-1}} \{(z-a)^m f(z)\}.$$

Especially,

$$\operatorname{Res}(f, a) = g(a) = \lim_{z \to a} (z-a) f(z)$$

when $m = 1$.

Theorem 3.11 ***(Residue Theorem)*** *If $f(z)$ is holomorphic on a region $U \subseteq \mathbb{C}$ except on the points $z_1, z_2, \cdots, z_n$, $f(z)$ is continuous on $\bar{U}$*

except on the points $z_1, z_2, \cdots, z_n$, and the boundary ∂U is a rectifiable simple curve, then

$$\int_{\partial U} f(z)\,dz = 2\pi i \sum_{k=1}^{n} \operatorname{Res}(f, z_k).$$

Theorem 3.12 ***(Residue Theorem)*** *If $f(z)$ is holomorphic on $\mathbb{C}^*$ except on the points $z_1, z_2, \cdots, z_n, \infty$ and these points are isolated singularities of $f(z)$, then the sum of residues of $f(z)$ at these isolated singularities is zero. That is,*

$$\sum_{k=1}^{n} \operatorname{Res}(f, z_k) + \operatorname{Res}(f, \infty) = 0.$$

These two theorems can be easily proved by using the Cauchy Integral Theorem. We shall omit the proof here.

The Residue Theorem itself is very simple. What is important is that it can be used to calculate certain definite integrals of functions for which the antiderivative cannot be expressed in closed form. There are some techniques for calculating definite integrals by the Residue Theorem, through the choices of $f(z)$ and the choices of integral paths, etc. Here are three simple examples:

Example 3.1 Evaluate the integral

$$\int_{-\infty}^{\infty} \frac{dx}{(1+x^2)^{n+1}},$$

where n is a positive integer.

Solution Let $f(z) = 1/(1+z^2)^{n+1}$. Then $f(z)$ has the only isolated singularity $z = i$ on the upper half plane, and it is a pole with multiplicity $n+1$. Let U be the upper half disc $\{z \mid |z| < R, \operatorname{Im} z > 0\}$ (see Figure 4). then

$$\begin{aligned}
\operatorname{Res}(f, i) &= \frac{1}{n!}\frac{d^n}{dz^n}\left\{\frac{(z-i)^{n+1}}{(z^2+1)^{n+1}}\right\}\bigg|_{z=i} \\
&= \frac{1}{n!}\frac{d^n}{dz^n}\left\{\frac{1}{(z+i)^{n+1}}\right\}\bigg|_{z=i} \\
&= \frac{1}{n!}\frac{(-1)^n(n+1)(n+2)\cdots(2n)}{(2i)^{2n+1}} \\
&= \frac{1}{2i}\frac{(2n)!}{2^{2n}(n!)^2}.
\end{aligned}$$

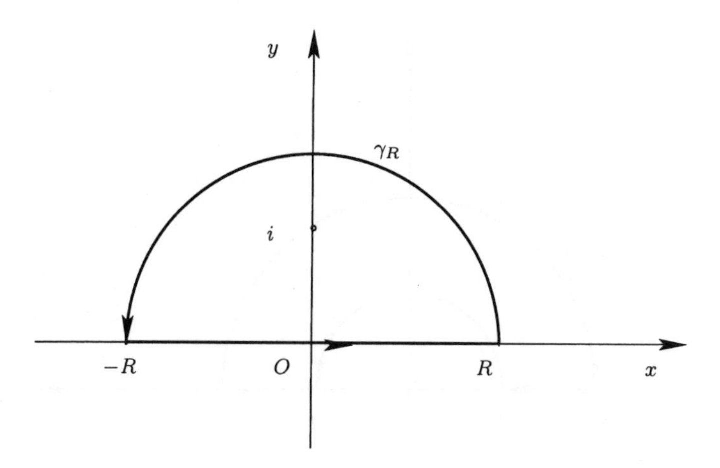

Fig. 4

On the other hand,

$$\begin{aligned}\operatorname{Res}(f,i) &= \frac{1}{2\pi i}\int_{\partial U}\frac{dz}{(1+z^2)^{n+1}} \\ &= \frac{1}{2\pi i}\int_{-R}^{R}\frac{dx}{(1+x^2)^{n+1}} + \frac{1}{2\pi i}\int_{\gamma_R}\frac{dz}{(1+z^2)^{n+1}},\end{aligned} \tag{3.14}$$

where γ_R is the upper half circle $\{z|\, z = Re^{i\theta}, 0 \le \theta \le \pi\}$ in Figure 4. The integral

$$\frac{1}{2\pi i}\int_{\gamma_R}\frac{dz}{(1+z^2)^{n+1}} = \frac{1}{2\pi i}\int_0^{\pi}\frac{iRe^{i\theta}d\theta}{(1+R^2e^{2i\theta})^{n+1}}$$

tends to zero as $R \to \infty$. Let $R \to \infty$ in (3.14), we get that

$$\int_{-\infty}^{\infty}\frac{dx}{(1+x^2)^{n+1}} = 2\pi i \operatorname{Res}(f,i) = \pi\frac{(2n)!}{2^{2n}(n!)^2}.$$

Example 3.2 Evaluate the integral (Dirichlet integral)

$$\int_0^{\infty}\frac{\sin x}{x}\,dx.$$

Solution Obviously,

$$\int_0^{\infty}\frac{\sin x}{x}\,dx = \frac{1}{2}\int_{-\infty}^{\infty}\frac{\sin x}{x}\,dx.$$

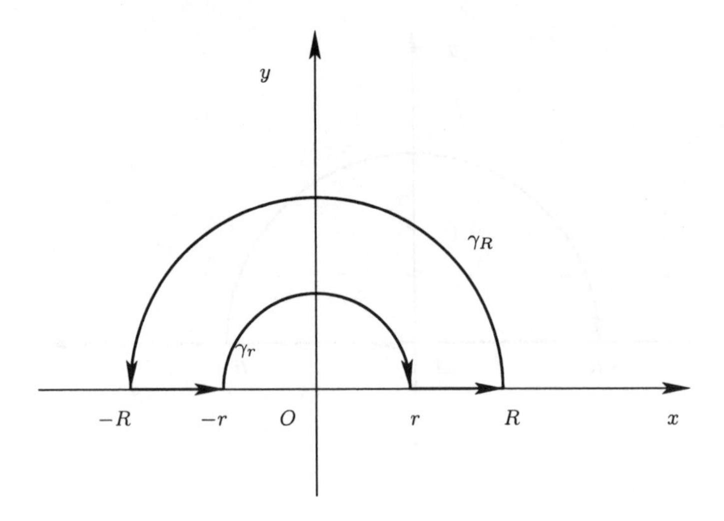

Fig. 5

Consider $f(z) = e^{iz}/z$. Let U be the half annulus on the upper half plane in Figure 5. Its boundary consists of:

$$-R < z < -r;\ r < z < R;$$

$$\gamma_r:\ z = re^{i\theta},\ 0 \leq \theta \leq \pi;$$

$$\gamma_R:\ z = Re^{i\theta},\ 0 \leq \theta \leq \pi.$$

Since $f(z)$ is holomorphic in this half annulus, by the Cauchy Integral Theorem we have

$$\int_r^R f(x)\,dx + \int_{-R}^{-r} f(x)\,dx + \int_{\gamma_R} f(z)\,dz + \int_{\gamma_r} f(z)\,dz = 0.$$

The third integral of the above equation is

$$\begin{aligned}\int_{\gamma_R} f(z)\,dz &= \int_0^\pi \frac{e^{iR(\cos\theta + i\sin\theta)}}{Re^{i\theta}} iRe^{i\theta}\,d\theta \\ &= i\int_0^\pi e^{-R\sin\theta + iR\cos\theta}\,d\theta.\end{aligned}$$

Hence

$$\left|\int_{\gamma_R} f(z)\,dz\right| \le \int_0^{\pi} e^{-R\sin\theta}\,d\theta = 2\int_0^{\frac{\pi}{2}} e^{-R\sin\theta}\,d\theta.$$

Since $(2/\pi)\theta \le \sin\theta$ for $0 \le \theta \le \pi/2$, we have

$$\int_0^{\frac{\pi}{2}} e^{-R\sin\theta}\,d\theta \le \int_0^{\frac{\pi}{2}} e^{-\frac{2R}{\pi}\theta}\,d\theta = \frac{\pi}{2R}(1-e^{-R}).$$

Since the right end tends to zero as $R \to \infty$, it follows that

$$\int_{\gamma_R} f(z)\,dz \to 0$$

as $R \to \infty$.

On the other hand,

$$\begin{aligned}\int_{\gamma_r} f(z)\,dz &= i\int_{\pi}^{0} e^{-r\sin\theta + ir\cos\theta}\,d\theta \\ &= i\int_{\pi}^{0}(1+O(r))\,d\theta = -\pi i + O(r).\end{aligned}$$

Thus

$$\int_{\gamma_r} f(z)\,dz \to -\pi i$$

as $r \to 0$.

Since

$$\int_{-R}^{-r} f(x)\,dx = \int_{-R}^{-r}\frac{e^{ix}}{x}\,dx = -\int_r^R \frac{e^{-ix}}{x}\,dx,$$

by letting $r \to 0$ and $R \to \infty$ in the equation

$$\int_r^R \frac{e^{ix}-e^{-ix}}{x}\,dx + \int_{\gamma_R} f(z)\,dz + \int_{\gamma_r} f(z)\,dz = 0,$$

we get

$$\int_0^{\infty}\frac{\sin x}{x}\,dx = \frac{\pi}{2}.$$

Example 3.3 Evaluate

$$\int_0^{\infty}\cos x^2\,dx, \quad \int_0^{\infty}\sin x^2\,dx.$$

Solution Let $f(z) = e^{iz^2}$, U be a sector region bounded by

$$I: \ 0 \le z \le R; \ \ II: \ re^{i\frac{\pi}{4}}, 0 \le r \le R$$

and

$$\gamma_R: \ Re^{i\theta}, 0 \le \theta \le \frac{\pi}{4}$$

(see Figure 6). Then by the Cauchy Integral Theorem, we have

$$\int_I f(z)\,dz + \int_{II} f(z)\,dz + \int_{\gamma_R} f(z)\,dz = 0.$$

The third integral of the above equation is

$$\int_{\gamma_R} f(z)\,dz = \int_0^{\frac{\pi}{4}} e^{iR^2(\cos 2\theta + i\sin 2\theta)} iRe^{i\theta}\,d\theta.$$

Thus

$$\left|\int_{\gamma_R} f(z)\,dz\right| \le R\int_0^{\frac{\pi}{4}} e^{-R^2(\sin 2\theta)}\,d\theta \le R\int_0^{\frac{\pi}{4}} e^{-R^2\frac{4}{\pi}\theta}\,d\theta = \frac{\pi}{4R}(1 - e^{-R^2}).$$

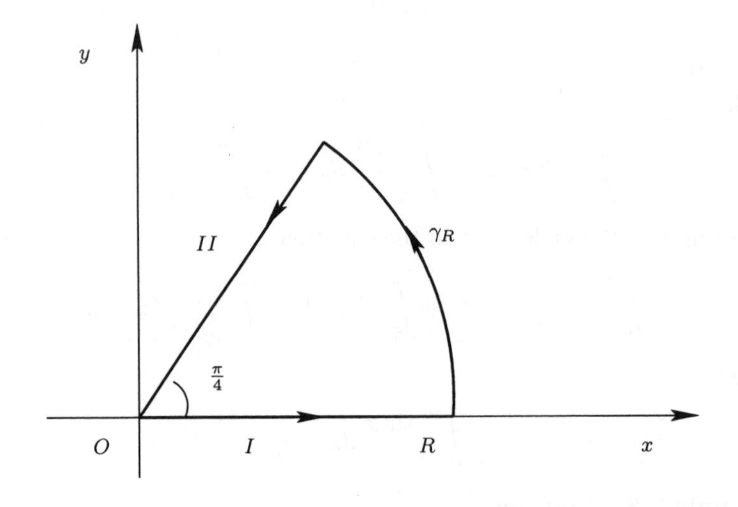

Fig. 6

Since the term at the right end tends to zero as $R \to \infty$, we have

$$\lim_{R\to\infty}\left(\int_I f(z)\,dz + \int_{II} f(z)\,dz\right) = 0.$$

That is

$$\int_0^\infty e^{ix^2}\,dx - \int_0^\infty e^{ix^2 e^{i\frac{\pi}{2}}} e^{\frac{\pi i}{4}}\,dx = 0.$$

It follows that

$$\int_0^\infty e^{ix^2}\,dx = e^{\frac{\pi}{4}i}\int_0^\infty e^{-x^2}\,dx.$$

Since

$$\int_0^\infty e^{-x^2}\,dx = \frac{\sqrt{\pi}}{2},$$

we have

$$\int_0^\infty e^{ix^2}\,dx = \frac{\sqrt{\pi}}{2}e^{\frac{\pi i}{4}}.$$

Therefore,

$$\int_0^\infty \cos x^2\,dx = \int_0^\infty \sin x^2\,dx = \frac{\sqrt{2\pi}}{4}.$$

3.6 Analytic Continuation

Let $f(z)$ be a holomorphic function on a region $U \subseteq \mathbb{C}$, G be a region that contains U and $F(z)$ be a holomorphic function on G. If $f(z) = F(z)$ on U, then we say that $F(z)$ is the *analytic continuation* of $f(z)$ to $G \setminus U$. By the theorem of the uniqueness of holomorphic functions, if such an F exists on G, then F is unique. Similarly, Let $f_1(z)$, $f_2(z)$ be holomorphic functions on regions U_1, U_2 respectively, where $U_1 \bigcap U_2 = U_3 \neq \emptyset$ and $f_1 = f_2$ on U_3. Define

$$f(z) = \begin{cases} f_1(z), & \text{if } z \in U_1, \\ f_2(z), & \text{if } z \in U_2 \end{cases}$$

on $U = U_1 \bigcup U_2$ (Figure 7). Then f is holomorphic on U and f_1, f_2 are called analytical continuations of each other.

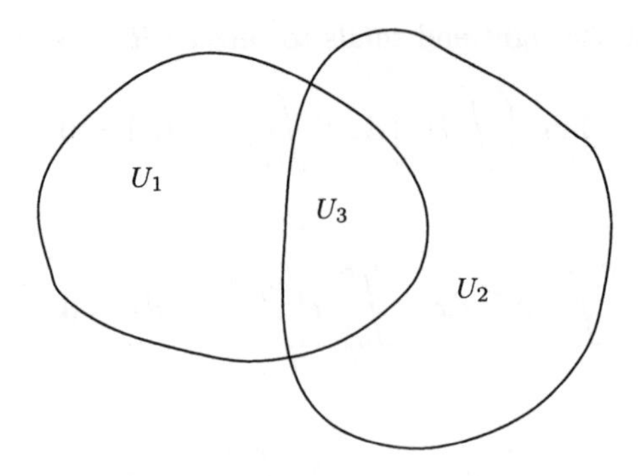

Fig. 7

The most important and natural way to obtain the analytical continuation of f is through the power series. According to the Abel Theorem (Theorem 1.14 in Section 1.6), for a power series

$$a_0 + a_1 z + \cdots + a_n z^n + \cdots, \tag{3.15}$$

there exists a radius of convergence R, such that the above series converges absolutely in $|z| < R$ and uniformly on every compact subset of $|z| < R$. Let $f(z) = \sum_{n=0}^{\infty} a_n z^n$. Then $f(z)$ is a holomorphic function. If $z_0 \in D(0, R)$, then $f(z)$ has a Taylor expansion at $z = z_0$,

$$f(z) = \sum_{n=0}^{\infty} \frac{f^n(z_0)}{n!}(z - z_0)^n.$$

Let ρ be the radius of convergence of this series. Then $\rho \geq R - |z_0|$. If $\rho > R - |z_0|$, then a part of $D(z_0, \rho)$ is outside $D(0, R)$. Thus $f(z)$ can be analytically continued to $D(z_0, \rho) \setminus D(0, R)$. If $\rho = R - |z_0|$, then $D(z_0, \rho)$ is tangent to $D(0, R)$ at a point ζ_0. It follows that $f(z)$ cannot be analytically continued from ζ_0. The point ζ_0 is called a *singularity* of $f(z)$.

We claim that, if R is the radius of convergence of (3.15), then there exists at least one singularity of $f(z)$ on $|z| = R$.

To prove this claim, let us assume the contrary. Then $f(z)$ can be analytically continued from any point of $|z| = R$. That is, for any point ζ on $|z| = R$, there exist $D(\zeta, r_\zeta)$ and $g_\zeta(z)$, where $g_\zeta(z)$ is holomorphic on $D(\zeta, r_\zeta)$ and $g_\zeta(z) = f(z)$ for $z \in D(0, R) \bigcap D(\zeta, r_\zeta)$. Since $|z| = R$ is compact, by the Heine-Borel Theorem, there exists a finite subcovering of $|z| = R$, $D(\zeta_1, r_{\zeta_1})$, $D(\zeta_2, r_{\zeta_2})$, $\cdots$, $D(\zeta_m, r_{\zeta_m})$, in $\{D(\zeta, r_\zeta)\}$. Let

$G = \bigcup_{k=1}^{m} D(\zeta_k, r_{\zeta_k})$ and ρ be the distance from $|z| = R$ to ∂G. Obviously, $\rho > 0$. Thus $R - \rho < |z| < R + \rho \subset G$. For $z \in D(\zeta_k, r_{\zeta_k})$ $(k = 1, 2, \cdots, m)$, define $\Phi(z) = g_{\zeta_k}(z)$ in G, then $\Phi(z)$ is a single valued holomorphic function on G. Indeed, if $D(\zeta_k, r_{\zeta_k}) \bigcap D(\zeta_l, r_{\zeta_l}) \neq \emptyset$, $k \neq l$, then $(D(\zeta_k, r_{\zeta_k}) \bigcap D(\zeta_l, r_{\zeta_l})) \bigcap D(0, R) \neq \emptyset$ and $g_{\zeta_k}(z) = g_{\zeta_l}(z) = f(z)$ on this intersection. By the uniqueness theorem of holomorphic functions, $g_{\zeta_k}(z) = g_{\zeta_l}(z)$ on $D(\zeta_k, r_{\zeta_k}) \bigcap D(\zeta_l, r_{\zeta_l})$. Since $\Phi(z) = f(z)$ on $G \bigcap D(0, R)$, $f(z)$ can be analytically continued to $G \bigcup D$. Since $D(0, R+\rho)$ is contained in $G \bigcup D$, this contradicts the assumption that the radius of convergence of (3.15) is R.

Therefore, (3.15) must has singularities on its circle of convergence. For instance, $z = 1$ is a singularity of $\sum_{n=0}^{N} z^n$. The next famous example tells us that there exists such a power series that every point on its circle of convergence is a singularity.

Example 3.4 Let

$$f(z) = z^{1!} + z^{2!} + \cdots + z^{n!} + \cdots . \tag{3.16}$$

Then every point on its circle of convergence is a singularity.

Proof Since

$$a_n = \begin{cases} 1, & \text{if } n = k!, \\ 0, & \text{if } n \neq k!, \end{cases}$$

we have

$$\overline{\lim_{n \to \infty}} \sqrt[n]{|a_n|} = 1$$

Thus, the radius of convergence of (3.16) is $R = 1$. That is $f(z)$ is holomorphic on $D(0, 1)$. If $z_0 \in D(0, 1)$ and $|z_0| = 1/2$, then $f(z)$ has a Taylor expansion

$$g(z) = \sum_{n=0}^{\infty} \frac{f^{(n)}(z_0)}{n!} (z - z_0)^n.$$

Extend Oz_0 to intersect with $|z| = 1$ at a point ζ_0. If we can show that the radius of convergence of $g(z)$ is $1/2$, then $f(z)$ cannot be analytically continued from ζ_0. Suppose the contrary. If ρ is the radius of convergence of $g(z)$ and $\rho > 1/2$, then $D(\zeta_0, \rho) \bigcap D(0, 1) \neq \emptyset$. Thus, $D(\zeta_0, \rho)$ contains an arc σ of $|z| = 1$ where $\zeta_0 \in \sigma$. Since the set of points $\exp\{(2\pi i p)/q\}$ (where p, q are integers and p/q is a reduced fraction) is dense on $|z| = 1$, there exists a point $\zeta_1 = \exp\{(2\pi ip)/q\}$ on σ and $\lim_{r\to 1} g(r\zeta_1) = g(\zeta_1)$, $(0 < r < 1)$. Since $g(z) = f(z)$ for $z \in D(0,1)$, we have $\lim_{r\to 1} f(r\zeta_1) = g(\zeta_1)$. Since

$$f(r\zeta_1) = \sum_{n=1}^{q-1} r^{n!}\xi_1^{n!} + \sum_{n=q}^{\infty} r^{n!},$$

it is easy to see that

$$\sum_{n=q}^{\infty} r^{n!} > \sum_{n=q}^{N} r^{n!} > (N-q)r^{N!}.$$

Thus, $\sum_{n=q}^{\infty} r^{n!}$ is greater then any positive integers as $r \to 1$. Therefore $\lim_{r\to 1} |f(r\zeta_1)| = \infty$ and this contradicts with $\lim_{r\to 1} f(r\zeta_1) = g(\zeta_1)$. This proves that ζ_0 is a singularity of $f(z)$.

A function is analytic (or holomorphic) near a point z_0 if it has a convergent power series expansion near the point. This is the definition of *locally analytic* (or *locally holomorphic*) functions. This definition is equivalent to the definition in Section 1.3 of Chapter 1. Using the concept of analytic continuation, we can define global analytic (or global holomorphic) functions.

Consider a locally holomorphic function

$$f(z) = \sum_{n=0}^{\infty} c_n (z-a)^n$$

and let R be its radius of convergence. If $a_1 \in D(a,R)$, then we have another power series

$$f_1(z) = \sum_{n=0}^{\infty} \frac{f^{(n)}(a_1)}{n!}(z-a_1)^n$$

with radius of convergence $R_1 \geq R - |a - a_1| > 0$. The holomorphic functions $f(z)$ and $f_1(z)$ are called *analytic elements* and $f_1(z)$ is an analytic continuation of $f(z)$. If we have m analytic elements

$$f_k(z) = \sum_{n=0}^{\infty} c_n^{(k)}(z-a_k)^n, \ (k = 1, 2, \cdots)$$

and f_k is an analytic continuation of f_{k-1}, then f_m is also an analytic continuation of f. The set of all the analytic elements obtained by analytic continuation of $f(z)$ is called a *global analytic function* (or a *global holomorphic function*).

A *complete analytic function* is a global analytic function which contains all analytic continuations of any of its analytic elements. In general, this is a multi-valued function. The union of discs of convergence of all analytic elements of a complete analytic function is called its *existence domain*. It obviously can not be continued further. Hence all points on the boundary of the existence domain are singularities of complete analytic functions.

Exercise III

1. Let $a_1, a_2, \cdots$ be a sequence of distinct points and $\lim_{n\to\infty} |a_n| = \infty$. If

$$\psi_n(z) = \sum_{j=1}^{\infty} \frac{c_{n,j}}{(z-a_n)^j}, \quad n = 1, 2, \cdots$$

are holomorphic on $\mathbb{C}$ except at points a_n $(n = 1, 2, \cdots)$, then there exists a holomorphic function $f(z)$ on $\mathbb{C}\backslash\{a_1, a_2, \cdots\}$ such that the principle parts of the Laurent expansion of $f(z)$ at a_n $(n = 1, 2, \cdots)$ are $\psi_n(z)$ $(n = 1, 2, \cdots)$ respectively.

2. Prove Theorem 3.11 and Theorem 3.12.

3. Find the Laurent series of the following functions on the indicated regions.

(i) $\dfrac{1}{z^3(z+i)}$, $\quad 0 < |z+i| < 1$;

(ii) $\dfrac{z^2}{(z+1)(z+2)}$, $\quad 1 < |z| < 2$;

(iii) $\log\left(\dfrac{z-a}{z-b}\right)$, $\quad \max(|a|, |b|) < |z| < \infty$;

(iv) $z^2 e^{\frac{1}{z}}$, $\quad 0 < |z| < \infty$;

(v) $\sin\dfrac{z}{1+z}$, $\quad 0 < |z+1| < \infty$.

4. Find and classify the singularities of the following functions. Find the order if the singularity is a pole.

(i) $\dfrac{\sin z}{z}$; $\quad$ (ii) $\dfrac{1}{z^2-1}\cos\dfrac{\pi z}{z+1}$; $\quad$ (iii) $z(e^{\frac{1}{z}} - 1)$;

(iv) $\sin\dfrac{1}{1-z}$; $\quad$ (v) $\dfrac{\exp\{\frac{1}{1-z}\}}{e^z - 1}$; $\quad$ (vi) $\tan z$

5. Prove: (1) If a is an essential singularity of $f(z)$ and $f(z) \neq 0$, then

a is also an essential singularity of $1/f(z)$.

(2) If a is an essential singularity of $f(z)$ and $P(\zeta)$ is a non-constant polynomial, then a is also an essential singularity of $P(f(z))$.

6. Prove:

(i) $\prod_{n=0}^{\infty}(1+z^{2^n}) = \dfrac{1}{1-z}$;

(ii) $\sinh \pi z = \pi z \prod_{n=1}^{\infty}\left(1+\dfrac{z^2}{n^2}\right)$;

(iii) $\cos \pi z = \prod_{n=0}^{\infty}\left(1-\left(\dfrac{z}{n+\dfrac{1}{2}}\right)^2\right)$;

(iv) $e^z - 1 = ze^{\frac{z}{2}}\prod_{n=1}^{\infty}\left(1+\dfrac{z}{4\pi^2 n^2}\right)$.

(v) $\sin \pi z = \pi z \prod_{n\neq 0}\left(1-\dfrac{z}{n}\right)e^{\frac{z}{n}}$, hence $\sin \pi z = \prod_{n=1}^{\infty}\left(1-\dfrac{z^2}{n^2}\right)$.

7. (*Blaschke Product*) Suppose a sequence of complex numbers $\{a_k\}$ satisfies $0 < |a_k| < 1$ $(k = 1, 2, \cdots)$ and $\sum_{k=1}^{\infty}(1-|a_k|) < \infty$. Show that the infinite product

$$f(z) = \prod_{k=1}^{\infty}\frac{a_k - z}{1-\bar{a_k}z}\cdot\frac{|a_n|}{a_n}$$

converges uniformly on the disc $\{z\,|\,|z| \le r, (0 < r < 1)\}$, and $f(z)$ is holomorphic in $|z| < 1$ with $|f(z)| \le 1$. Also show that the zeros of $f(z)$ are $a_k (k = 1, 2, \cdots)$ and there is no other zeros.

8. (*Poisson-Jensen Formula*) Suppose $f(z)$ is a meromorphic function on the disc $\{z\,|\,|z| \le R\}$, $(0 < R < \infty)$, and is not equal to zero identically, the points $a_1, a_2, \cdots, a_s$ and $b_1, b_2, \cdots, b_t$ are zeros and poles of $f(z)$ on $|z| < R$ respectively. Suppose that if a is a zero with multiplicity n, then a appears n times in $a_1, a_2, \cdots, a_s$. Same assumption for the poles. Then we have

$$\log|f(z)| = \frac{1}{2\pi}\int_0^{2\pi}\log|f(Re^{i\varphi})|\,\mathrm{Re}\,\frac{Re^{i\varphi}+2}{Re^{i\varphi}-z}\,d\varphi$$
$$+\sum_{j=1}^{t}\log\left|\frac{R^2-\bar{b_j}z}{R(z-b_j)}\right| - \sum_{i=1}^{s}\log\left|\frac{R^2-\bar{a}_i z}{R(z-a_i)}\right|$$

for any points in the disc $|z| < 1$ which is different from a_i $(i = 1, 2, \cdots, s)$

and b_j $(j = 1, 2, \cdots, t)$.

This formula is the starting point of Nevanlinna Value Distribution theory.

9. Suppose a meromorphic function $f(z)$ only has two poles on the extended complex plane $\mathbb{C}^*$. The point $z = -1$ is a pole of multiplicity 1 and with principle part $1/(z+1)$; the point $z = 2$ is a pole of multiplicity 2 and with principle part $2/(z-2) + 3/(z-2)^2$, and $f(0) = 7/4$. Find the Laurent expansion of $f(z)$ in $1 < |z| < 2$.

10. Suppose a meromorphic function $f(z)$ has poles at $z = 1, 2, 3, \cdots$ with multiplicity 2, and the principle parts of Laurent expansions in the neighborhood of $z = n$ are $\psi_n(z) = n/(z-n)^2$ $(n = 1, 2, \cdots)$. Find a general form of $f(z)$.

11. (i) Expand $f(z) = 1/(e^z - 1)$ to a partial fraction.

(ii) Show that

$$\frac{1}{\sin^2 \pi z} = \frac{1}{\pi^2} \sum_{n=-\infty}^{\infty} \frac{1}{(z-n)^2};$$

(iii) Show that if $\alpha \neq 0$ and $(\beta/\alpha) \neq \pm 1, \pm 2, \cdots$, then

$$\frac{\pi}{\alpha} \cot \frac{\pi\beta}{\alpha} = \sum_{n=0}^{\infty} \left\{ \frac{1}{n\alpha + \beta} - \frac{1}{n\alpha + (\alpha - \beta)} \right\}.$$

From the above equation show that

$$\frac{1}{1 \cdot 2} + \frac{1}{4 \cdot 5} + \frac{1}{7 \cdot 8} + \cdots + \frac{1}{(3n-2)(3n-1)} + \cdots = \frac{\pi}{3\sqrt{3}}.$$

12. Suppose a meromorphic function $f(z)$ only has finitely many poles $\alpha_1, \alpha_2, \cdots, \alpha_m$ and α_k is not a integer $(1 \leq k \leq m)$. The infinity $z = \infty$ is a zero of $f(z)$ with multiplicity $p \geq 2$. Show that

(1) $\displaystyle\lim_{n\to\infty} \sum_{k=-n}^{n} (-1)^k f(k) = -\pi \sum_{k=1}^{m} \operatorname{Res}(f(z) \cot \pi z, \alpha_k);$

(2) $\displaystyle\lim_{n\to\infty} \sum_{k=-n}^{n} (-1)^k f(k) = -\pi \sum_{k=1}^{m} \operatorname{Res}\left(\frac{f(z)}{\sin \pi z}, \alpha_k\right);$

(3) Find the sums of the following series by using (1) and (2).

(i) $\displaystyle\sum_{n=-\infty}^{\infty} \frac{1}{(a+n)^2}$, where a is not an integer;

(ii) $\displaystyle\sum_{n=0}^{\infty} \frac{(-1)^n}{n^2 + a^2}$, where a is a non-zero real number.

13. Find the residues of the following functions at their isolated singularities (including the infinity point if it is an isolated singularity).

(i) $\dfrac{1}{z^2 - z^4}$; (ii) $\dfrac{z^2 + z + 2}{z(z^2 + 1)^2}$;

(iii) $\dfrac{z^{n-1}}{z^n + a^n}$, where $a \neq 0$ and n is a positive integer;

(iv) $\dfrac{1}{\sin z}$; (v) $z^3 \cos \dfrac{1}{z-2}$; (vi) $\dfrac{e^z}{z(z+1)}$.

14. Suppose that $f(z)$ and $g(z)$ are holomorphic at $z = a$, $f(a) \neq 0$ and $z = a$ is a zero of $g(z)$ with multiplicity 2. Find $\text{Res}(f(z)/g(z), a)$.

15. Evaluate the following integrals:

(i) $\displaystyle\int_0^{+\infty} \frac{x^2 dx}{(x^2+1)^2}$; (ii) $\displaystyle\int_0^{\frac{\pi}{2}} \frac{dx}{a + \sin^2 x}$, $a > 0$;

(iii) $\displaystyle\int_0^{+\infty} \frac{x \sin x}{x^2 + 1} dx$; (iv) $\displaystyle\int_0^{+\infty} \frac{\log x}{(x^2+1)^2} dx$;

(v) $\displaystyle\int_0^{+\infty} \frac{x^{1-\alpha}}{1+x^2} dx$, $0 < \alpha < 2$;

(vi) $\displaystyle\int_0^{+\infty} \frac{dx}{1+x^n}$, where n is an integer and $n \geq 2$;

(vii) $\displaystyle\int_0^{\pi} \frac{d\theta}{a + \cos\theta}$, where a is a constant and $a > 1$;

(viii) $\displaystyle\int_0^{+\infty} \left(\frac{\sin x}{x}\right)^2 dx$;

(ix) $\displaystyle\int_0^{+\infty} \frac{x^p}{x^2 + 2x\cos\lambda + 1} dx$, where $-1 < p < 1$ and $-\pi < \lambda < \pi$;

(x) $\displaystyle\int_0^{+\infty} \frac{1}{1+x^p} dx$, where $p > 1$;

(xi) $\displaystyle\int_0^{1} \frac{x^{1-p}(1-x)^p}{1+x^2} dx$, where $-1 < p < 2$;

(xii) $\displaystyle\int_0^{+\infty} \frac{\log x}{x^2 + 2x + 2} dx$;

(xiii) $\displaystyle\int_0^{+\infty} \frac{\sqrt{x}\log x}{x^2+1} dx$;

(xiv) $\displaystyle\int_0^{+\infty} \log\left(\frac{e^x+1}{e^x-1}\right) dx$;

(xv) $\displaystyle\int_0^{+\infty} \frac{x}{e^x+1} dx$;

(xvi) $\displaystyle\int_0^{\frac{\pi}{2}} \log\sin\theta \, d\theta$;

(xvii) $\displaystyle\int_0^{\pi} \frac{x \sin x}{1 - 2a \cos x + a^2}\, dx$, where $a > 0$;

(xviii) $\displaystyle\int_0^{+\infty} \frac{\log(1 + x^2)}{1 + x^2}\, dx$.

16. Explain whether the function defined by the series

$$-\frac{1}{z} - 1 - z - z^2 - \cdots$$

in $0 < |z| < 1$ can be analytically continued to the function defined by the series

$$\frac{1}{z^2} + \frac{1}{z^3} + \frac{1}{z^4} + \cdots$$

in $|z| > 1$.

17. Show that the function defined by the series

$$1 + \alpha z + \alpha^2 z^2 + \cdots + \alpha^n z^n + \cdots$$

and the function defined by

$$\frac{1}{1-z} - \frac{(1-\alpha)z}{(1-z)^2} + \frac{(1-\alpha)^2 z^2}{(1-z)^3} - \cdots$$

are analytic continuations of each other.

18. Show that the function $f_1(z)$ defined by the series

$$z - \frac{1}{2}z^2 + \frac{1}{3}z^3 - \cdots$$

in $|z| < 1$, and the function $f_2(z)$ defined by the series

$$\ln 2 - \frac{1-z}{2} - \frac{(1-z)^2}{2 \cdot 2^2} - \frac{(1-z)^3}{3 \cdot 2^3} - \cdots$$

in $|z - 1| < 2$ are analytic continuations of each other.

19. Prove that the power series $\sum_{n=0}^{\infty} z^{2^n}$ cannot be analytically continued to the outside of its circle of convergence.

Chapter 4

Riemann Mapping Theorem

4.1 Conformal Mapping

Another important part of complex analysis is the theory of Riemann conformal mapping. The basic idea of this theory is to treat holomorphic functions from the geometric point of view. By Section 1.3, if $f'(z) \neq 0$, then the map $w = f(z)$ is conformal and is called a *conformal mapping* or a *holomorphic mapping*.

Note that if $U \subseteq \mathbb{C}$ is a region, $w = f(z)$ is a holomorphic map, then $f(U)$ is also a region.

Indeed, if w_1 and w_2 are two arbitrary points in $f(U)$, then there exist z_1 and z_2 in U such that $w_1 = f(z_1)$ and $w_2 = f(z_2)$. Since U is connected, there exists $\gamma(t)$ join z_1 and z_2 in U. It follows that $f(\gamma(t)) \subset f(U)$ joins w_1 and w_2. Thus $f(U)$ is connected.

Suppose that w_0 is an arbitrary point in $f(U)$. By Theorem 2.17 in Section 2.4, for a sufficiently small $\rho > 0$, there exists a $\delta > 0$, such that for any point w in $D(w_0, \delta)$, there exists a point z in $D(z_0, \rho)$ with $f(z) = w$. It follows that $D(w_0, \delta) \subset f(U)$ and $f(U)$ is an open set. This assertion is also referred to as the Open Mapping Theorem: f maps open sets to open sets.

It was defined in Section 1.5 that a function $f(z)$ on a region $U \subseteq \mathbb{C}$ is univalent if for any two points z_1 and z_2 in U, $f(z_1) = f(z_2)$ if and only if $z_1 = z_2$. If $f(z)$ is univalent and holomorphic on $U \subseteq \mathbb{C}$, then $f'(z) \neq 0$ for any point $z \in U$. Conversely, if $f'(z_0) \neq 0$ for $z_0 \in U$, then $f(z)$ is univalent in a neighborhood of z_0.

To prove the first assertion, suppose the contrary, $f(z)$ is univalent and holomorphic on a region $U \subseteq \mathbb{C}$, but there exists $z_0 \in U$ such that $f'(z_0) = 0$. Then z_0 is a zero of the function $f(z) - f(z_0)$ with multiplicity $m \geq 2$.

Choose a neighborhood of z_0 such that $f'(z) \neq 0$ in this neighborhood except at the point z_0. By Theorem 2.17 in Section 2.4, $f(z) - w$ has exact m zeros in the chosen neighborhood of z_0 for any w in the corresponding neighborhood of $w_0 = f(z_0)$. This contradicts with the assumption that $f(z)$ is univalent on U. Conversely, if $f'(z_0) \neq 0$, then z_0 is a simple zero of $f(z) - f(z_0)$. By Theorem 2.17 in Section 2.4, for any $\rho > 0$ sufficiently small, there exists a $\delta > 0$, such that $f(z) - w$ has only one zero in $D(z_0, \rho)$ where w is an arbitrary point in $D(f(z_0), \delta)$. In other words, there is only one z such that $f(z) = w$. Choose $\rho_1 < \rho$ such that $f(D(z_0, \rho_1)) \subset D(w_0, \delta)$. Then $f(z)$ is univalent on $D(z_0, \rho_1)$.

Moreover, it is easy to see that if $w = f(z)$ is univalent and holomorphic on U and maps U to G, then its inverse $z = g(w)$ is univalent and holomorphic on G and maps G to U. So the univalent holomorphic mapping is also called *biholomorphic mapping*.

Theorem 4.1 *Let $G \subseteq \mathbb{C}$ be a region, γ be a rectifiable simple closed curve in G, $U \subset G$ be the region inside γ. If $f(z)$ is holomorphic on G and maps γ one-to-one to a simple closed curve Γ, then $w = f(z)$ is univalent on U and maps U to the inside region V of Γ.*

Proof Suppose w_0 is not on Γ. By the Argument Principle (Theorem 2.14 of Section 2.4), the number of zeros N of $f(z) - w_0$ inside γ is

$$\frac{1}{2\pi i}\int_\gamma \frac{f'(z)}{f(z) - w_0}\,dz = \pm\frac{1}{2\pi i}\int_\Gamma \frac{dw}{w - w_0}.$$

If w_0 lies outside of Γ, then

$$\int_\Gamma \frac{dw}{w - w_0} = 0$$

Hence $N = 0$ and $f(z) - w$ has no zero in U. If w_0 lies inside of Γ, then

$$\frac{1}{2\pi i}\int_\Gamma \frac{dw}{w - w_0} = 1.$$

Thus

$$\frac{1}{2\pi i}\int_\gamma \frac{f'(z)}{f(z) - w_0}\,dz = \pm 1.$$

Since N is non-negative, we have $N = 1$ and $f(z) - w_0$ has only one zero in U. When z travels once along γ in a positive direction, $w = f(z)$ also travels once along Γ in a positive direction. We claim that when $w_0 \in \Gamma$, $f(z) - w_0$ has no zero in U. To see this, suppose the contrary, then there exists $z_0 \in U$ such that $f(z_0) = w_0$. Thus $D(w_0, \delta) \subset f(U)$. For every point $w_1 \in D(w_0, \delta)$, $f(z) - w_1$ has zeros in U. For $w_1 \in D(w_0, \delta)$ outside of Γ, this contradicts with the assertion $f(z) - w_1$ has no zero in U.

Here are some examples of conformal mapping.

Example 4.1 The only univalent holomorphic mapping which maps the unit disc $D(0,1)$ to itself are

$$w = e^{i\theta}\frac{z-a}{1-\bar{a}z}, \quad a \in D(0,1), \quad \theta \in \mathbb{R}. \tag{4.1}$$

Example 4.2 The only univalent holomorphic mappings which map the upper half plane $\operatorname{Im} z > 0$ to the unit disc $D(0,1)$ are

$$w = e^{i\theta}\frac{z-a}{z-\bar{a}}, \quad \operatorname{Im} a > 0, \quad \theta \in \mathbb{R}. \tag{4.2}$$

It is easy to see that this transformation maps $\operatorname{Im} z = 0$ to $|w| = 1$. So (4.2) can also be written as

$$z = \frac{\bar{a}w - e^{i\theta}a}{w - e^{i\theta}}$$

which maps $|w| = 1$ to $\operatorname{Im} z = 0$. By Theorem 4.1, this is a univalent holomorphic map from $D(0,1)$ to $\operatorname{Im} z > 0$. Conversely, if $w = f(z)$ is a univalent holomorphic map from $\operatorname{Im} z > 0$ to $D(0,1)$, then $f \circ \psi^{-1}$ maps the unit disc to itself. By Example 1, $f \circ \psi^{-1}$ must has the form of (4.1). Let $\varphi = f \circ \psi^{-1}$. Then $f = \varphi \circ \psi$ and it has the form of (4.2).

Similarly, we have

Example 4.3 The only univalent holomorphic functions that map the upper half plan $\operatorname{Im} z > 0$ to the upper half plane $\operatorname{Im} w > 0$ are

$$w = \frac{az+b}{cz+d},$$

where a, b, c, d are real numbers and $ad - bc > 0$.

It was proven in Section 3.3 that the group formed by all fractional linear transformations of the form $w = (az+b)/(cz+d)$, where $a, b, c, d \in \mathbb{C}$ and $ad - bc = 1$, is the group of meromorphic automorphisms $\operatorname{Aut}\{\mathbb{C}^*\}$. This group is isomorphic to the Möbius group $SL(2,\mathbb{C})/\{\pm I\}$ by the isomorphism

$$\frac{az+b}{cz+d} \longmapsto \begin{pmatrix} a & b \\ c & d \end{pmatrix} /\{\pm I\}.$$

Let $z = (aw+b)/(cw+d)$ be an arbitrary linear fractional transformation. If we consider a straight line as a circle with radius ∞, the important property of linear fractional transformation is that it maps a circle to a circle. The proof is as follows:

Let $z = x + iy$. Then the equation of a circle is

$$\alpha(x^2+y^2)+\beta x+\gamma y+\delta=0,$$

where α, β, γ and δ are real numbers. This equation can also be written as

$$\alpha z\bar{z}+\frac{1}{2}\beta(z+\bar{z})+\frac{1}{2i}\gamma(z-\bar{z})+\delta=0,$$

or

$$Az\bar{z}+Bz+\bar{B}\bar{z}+C=0, \tag{4.3}$$

where $A=\alpha, C=\delta$ are real numbers and $B=(1/2)\beta+(1/2i)\gamma$ is a complex number. If $\alpha=0$, then $A=0$, and (4.3) represents a straight line. Otherwise (4.3) represents a circle. Since every linear fractional transformation $z=(aw+b)/(cw+d)$ is a composition of translation $z=w+b$, multiplication $z=aw$ and inversion $z=1/w$, we only need to show that these three transformations map circles to circles. Substitute $z=w+b$ in (4.3), we have

$$Aw\bar{w}+(A\bar{b}+B)w+(Ab+\bar{B})\bar{w}+Ab\bar{b}+Bb+C=0,$$

and this equation is in the form of (4.3). Similarly, substitute $z=aw$ in (4.3), we have

$$Aa\bar{a}w\bar{w}+Baw+\bar{B}\bar{a}\bar{w}+C=0,$$

and substitute $z=1/w$ in (4.3), we have

$$Cw\bar{w}+\bar{B}w+B\bar{w}+A=0.$$

The last two equations are also of the form of (4.3) so the assertion follows.

Let z_1, z_2, z_3, z_4 be four points in $\mathbb{C}^*$ and at least three points are distinct. Define the *cross ratio* of these four points by

$$(z_1,z_2,z_3,z_4)=\frac{(z_1-z_3)(z_2-z_4)}{(z_1-z_4)(z_2-z_3)}.$$

If one of the z_i $(i = 1, 2, 3, 4)$ is ∞, then we define the cross ratio by limit. For instance

$$(\infty, z_2, z_3, z_4) = \frac{z_2 - z_4}{z_2 - z_3}.$$

Thus, we can show that if the linear fractional transformation $w = (az + b)/(cz + d)$ sends z_1, z_2, z_3, z_4 to w_1, w_2, w_3, w_4, then

$$(w_1, w_2, w_3, w_4) = (z_1, z_2, z_3, z_4)$$

In other words, the cross ratio is an invariant under linear fractional transformations. This assertion can be proved by substitute $w = (az+b)/(cz+d)$ in (w_1, w_2, w_3, w_4), to get (z_1, z_2, z_3, z_4).

To prove this assertion in another way, let $w_i = (az_i + b)/(cz_i + d)$, $i = 2, 3, 4$. Then $(z, z_2, z_3, z_4) = (w, w_2, w_3, w_4)$ is a Möbius transformation which send z_2 to w_2, z_3 to w_3 and z_4 to w_4. Since a Möbius transformation can be determined by three points, we have $w = (az + b)/(cz + d)$. Thus, if $w_1 = (az_1 + b)/(cz_1 + d)$ then $(w_1, w_2, w_3, w_4) = (z_1, z_2, z_3, z_4)$.

On the other hand, if there exists a function $f(z_1, z_2, z_3, z_4)$ which is an invariant under the group of linear fractional transformations, then f is a function of the cross ratio. In other words, the cross ratio is the only invariant under linear fractional transformations. In fact, by the assumption, the equation

$$f(Tz_1, Tz_2, Tz_3, Tz_4) = f(z_1, z_2, z_3, z_4)$$

holds for any linear fractional transformation T. Let $T_1 = z - z_4$. Then

$$f(z_1, z_2, z_3, z_4) = f(z_1 - z_4, z_2 - z_4, z_3 - z_4, 0).$$

Let $T_2 = 1/z$. Then

$$f(z_1, z_2, z_3, z_4) = f(\frac{1}{z_1 - z_4}, \frac{1}{z_2 - z_4}, \frac{1}{z_3 - z_4}, \infty).$$

Let $T_3 = z - 1/(z_3 - z_4)$. Then

$$f(z_1, z_2, z_3, z_4) = f\left(\frac{z_3 - z_1}{(z_1 - z_4)(z_3 - z_4)}, \frac{z_3 - z_2}{(z_2 - z_4)(z_3 - z_4)}, 0, \infty\right).$$

Let

$$T_4 = \frac{(z_2 - z_4)(z_3 - z_4)}{(z_3 - z_2)} z.$$

Then

$$f(z_1, z_2, z_3, z_4) = f\left(\frac{(z_3 - z_1)(z_2 - z_4)}{(z_1 - z_4)(z_3 - z_2)}, 1, 0, \infty\right)$$
$$= f((z_1, z_2, z_3, z_4), 1, 0, \infty).$$

This completes the proof.

4.2 Normal Family

The most important theorem in the Riemann conformal mapping theory is the Riemann Mapping Theorem.

Theorem 4.2 ***(Riemann Mapping Theorem)*** *Let $U \subseteq \mathbb{C}$ be a simply connected region whose boundary contains more than one point. Let z_0 be an arbitrary point in U. Then there exists only one univalent holomorphic function $f(z)$ on U that maps U to the unit disc $D(0,1)$ with $f(z_0) = 0$ and $f'(z_0) > 0$.*

It is natural that we require more than one point on the boundary. Suppose there is only one point on the boundary we may assume the point is ∞. If there exists a $f(z)$ which maps this region to the unit disc, then by Liouville Theorem (Theorem 2.10 in Section 2.3) $f(z)$ is a constant function.

Theorem 4.2 asserts that for any two simply connected regions whose boundaries contain more than one point, there exists a univalent holomorphic function that maps one to the other.

Let U and V be two regions in $\mathbb{C}$. If there exists a univalent holomorphic function that maps U to V, then we say that U and V are *holomorphically equivalent*. Thus Riemann Mapping Theorem tells us that any two simply connected regions whose boundaries contain more than one point are holomorphically equivalent. It is easy to see that any two simply connected regions are topologically equivalent. That is, there exists a continuous transformation which maps one to the other. Riemann Mapping Theorem also tells us that topological equivalence implies holomorphic equivalence. This is a very profound theorem. In Chapter 6, we will see that this theorem does not hold in higher dimensions (Theorem 6.11 in Section 6.4, Poincaré Theorem). This emphasized the special position of this theorem in the theory of complex analysis of one variable.

To prove this theorem, we need the concept of normal family of holomorphic functions. It is a basic concept in the function theory that is in a sense similar to compact sets in set theory. Further discussion about it can be found in Chapter 5.

Definition 4.1 A family of functions $\mathcal{F}$ is called *normal* on a region U if for any sequence in $\mathcal{F}$, there exists a subsequence that is uniformly convergent on every compact subset of U.

Theorem 4.3 *(Montel Theorem)* *Let $U \subseteq \mathbb{C}$ be a region, $\mathcal{F}$ be a family of holomorphic function on U. If there exists a positive number M such that*

$$|f(z)| \leq M$$

for all $z \in U$ and $f \in \mathcal{F}$, then $\mathcal{F}$ is normal.

In order to prove Theorem 4.3, we need to prove Ascoli-Arzela Theorem first.

Definition 4.2 Let $\mathcal{F} = \{f\}$ be a family of functions on a region $S \subseteq \mathbb{R}^n$. If for any $\varepsilon > 0$, there exists a $\delta > 0$ such that

$$|f(z) - f(w)| < \varepsilon$$

for any $f \in \mathcal{F}$ and any two points z and w in S which satisfy $|z - w| < \delta$, then $\mathcal{F}$ is called *equicontinuous*.

Definition 4.3 Let $\mathcal{F} = \{f\}$ be a family of function on a region $S \subseteq \mathbb{R}^n$. If there exists a positive number $M > 0$ such that for any $z \in S$ and $f \in \mathcal{F}$,

$$|f(z)| \leq M$$

holds, then $\mathcal{F}$ is called *equibounded*.

Theorem 4.4 *(Ascoli-Arzela Theorem)* *Let K be a compact set in $\mathbb{R}^n$. If a family of functions $\mathcal{F} = \{f_\nu\}$ is equicontinuous and equibounded, then there exists a subsequence that is uniformly convergent on K. In other words, on compact sets, equicontinuity and equiboundedness imply uniform convergence of a subsequence.*

Proof There exists a sequence $\{\zeta_k\}$ that is dense everywhere on K. For instance, the set of all points with rational coordinates. Since $\mathcal{F}$ is equibounded on K, for ζ_1, we can find a convergent subsequence $\{f_{\nu_{1k}}(\zeta_1)\}$ in $\{f_\nu(\zeta_1)\}$. We can also find a convergent subsequence $\{f_{\nu_{2k}}(\zeta_2)\}$ in $\{f_{\nu_{1k}}(\zeta_2)\}$, $\cdots$. Continuing this process, we get

$$\nu_{11} < \nu_{12} < \cdots < \nu_{1j} < \cdots,$$

$$\nu_{21} < \nu_{22} < \cdots < \nu_{2j} < \cdots ,$$

$$\cdots\cdots\cdots\cdots\cdots\cdots\cdots\cdots \tag{4.4}$$

$$\nu_{k1} < \nu_{k2} < \cdots < \nu_{kj} < \cdots ,$$

$$\cdots\cdots\cdots\cdots\cdots\cdots\cdots\cdots$$

where every row is a subsequence of the preceding row and $\lim_{j\to\infty} f_{\nu_{kj}}(\zeta_k)$ exists for every k. Obviously, $\{\nu_{jj}\}$ is a strictly increasing sequence and also a subsequence of every row of (4.4) except finite terms. Thus, $\{f_{\nu_{jj}}\}$ is a subsequence of $\{f_\nu\}$ that converges at every point ζ_k. Without loss of generosity, denote ν_{jj} by ν_j.

Since $\mathcal{F}$ is equicontinuous on K, for any given $\varepsilon > 0$, there exists a $\delta > 0$ such that for any $z, z' \in K$ and $f \in \mathcal{F}$, $|z - z"| < \delta$ implies $|f(z) - f(z')| < \varepsilon/3$. Since K is compact, it can be covered by finite many discs with radius $\delta/2$. Choose a point ζ_k in each disc, then there exists a N such that

$$|f_{\nu_i}(\zeta_k) - f_{\nu_j}(\zeta_k)| < \frac{\varepsilon}{3}$$

when $i, j > N$. For every $z \in K$, there exists a point ζ_k such that $|\zeta_k - z| < \delta$. Thus we have

$$|f_{\nu_i}(z) - f_{\nu_i}(\zeta_k)| < \frac{\varepsilon}{3}$$

and

$$|f_{\nu_j}(z) - f_{\nu_j}(\zeta_k)| < \frac{\varepsilon}{3}.$$

Therefore

$$|f_{\nu_i}(z) - f_{\nu_j}(z)| < \varepsilon$$

when $i, j > N$. Since K is compact, $\{f_{\nu_j}\}$ converges uniformly on K and the theorem follows.

Remark (1) The conditions of equicontinuity and equiboundedness in Theorem 4.4 are not only sufficient but also necessary for uniform convergence. We omit the proof here.

(2) Euclidean metric is used in the definition of equicontinuity as well as in Theorem 4.4. Theorem 4.4 also holds for other metrics (see Section 5.1 in Chapter 5). This point will be used in Theorem 5.7 of Section 5.5. We omit the proof here.

We now prove Montel Theorem (Theorem 4.3) by applying Ascoli-Arzela Theorem (Theorem 4.4).

Proof of Theorem 4.3 For any $z_0 \in U$, there exists a $R > 0$ such that $\bar{D}(z_0, R) \subseteq U$. Since U is an open set, the complement of U, U^c, is closed. The disc $\bar{D}(z_0, R)$ is disjoint from U^c. And the distance between these two closed sets is positive. In other words, there exists a $c > 0$ such that $|z - u| > c$ for $z \in \bar{D}(z_0, R)$ and $u \in U^c$. For any $z \in \bar{D}(z_0, R)$ and $f \in \mathcal{F}$, apply Cauchy inequality on $D(z, c)$, we get $|f'(z)| \leq M/c$. Let $C = M/c$. We have

$$|f(z) - f(w)| \leq C|z - w|$$

for any $z, w \in \bar{D}(z_0, R)$. This shows that $\mathcal{F}$ is equicontinuous on $\bar{D}(z_0, R)$. In fact, for any given $\varepsilon > 0$, choose $\delta = \varepsilon/C$ and the assertion follows.

If K is a compact subset of U, then there exist finitely many $\bar{D}(z_0, R)$ which can cover K. It follows that $\mathcal{F}$ is also equicontinuous on K. According to Ascoli-Arzela Theorem (Theorem 4.4), for any sequence $\{f_\nu\}$ in $\mathcal{F}$, there exists a subsequence $\{f_{\nu_k}\}$ that converges uniformly on K. Therefore, the same diagonal method which was used in the proof of Theorem 4.4, can also be used to show that there exists a subsequence $\{f_{\nu_k}\}$ of $\{f_\nu\}$ that converges uniformly on every compact subset of U. This proves the theorem.

In Chapter 5 Section 5.5, we will generalize the concept of normal family and the Montel Theorem to prove the famous Picard Theorem.

4.3 Riemann Mapping Theorem

In this section, we prove the Riemann Mapping Theorem (Theorem 4.2) by applying the Montel Theorem (Theorem 4.3).

Proof of Theorem 4.2 Let U be a bounded region, z_0 be a fixed point in U, $\mathcal{F} = \{\sigma(z)\}$ be the family of univalent holomorphic functions where $\sigma(z)$ maps U into $D(0, 1)$ and $\sigma(z_0) = 0$. The function family $\mathcal{F}$ is well defined and non-empty. Indeed, since U is bounded, there exists a $R > 0$ such that $U \subseteq D(0, R)$. The function $\sigma(\zeta) = (1/2R)(\zeta - z_0)$ maps z_0 to 0, it is univalent, holomorphic and satisfies $|\sigma(\zeta)| < (1/2R)(R + R) = 1$. This implies that $\sigma \in \mathcal{F}$ and $\mathcal{F}$ is non-empty. Since every function in $\mathcal{F}$ is holomorphic and bounded (the upper bound is 1), by the Montel Theorem

(Theorem 4.3), $\mathcal{F}$ is a normal family. Let

$$M = \sup\{|\sigma'(z_0)|\ |\sigma \in \mathcal{F}\}.$$

If $\bar{D}(z_0, r)$ is a closed disc with center z_0 and radius r contained in U, then $|\sigma'(z_0)| \leq 1/r$ by Cauchy inequality. Thus, $M \leq 1/r$. Now we show that there exists a $\sigma_0 \in \mathcal{F}$ such that $\sigma_0'(z_0) = M$.

By the definition of M, there exists a sequence $\{\sigma_j\}$ in $\mathcal{F}$ such that $|\sigma_j'(z_0)| \to M$. Since $\mathcal{F}$ is a normal family, there exists a subsequence $\{\sigma_{j_k}\}$ that converges uniformly to σ_0 on every compact subset of U. Since $|\sigma_{j_k}'(z_0)| \to M$, we have $|\sigma_0'(z_0)| = M$. When we multiply a complex number with norm 1 to σ_0, we get a new σ_0 such that $\sigma_0'(z_0) = M$.

Next, we prove that σ_0 is univalent on U by using the Argument Principle (Theorem 2.14 in Section 2.4). Let Q, R be two different points in U and $0 < s < |Q - R|$. Consider the function $\psi_k(z) = \sigma_{j_k}(z) - \sigma_{j_k}(Q)$ on $D(R, s)$. Since σ_j is univalent, ψ_k is not zero on $\bar{D}(R, s)$. By the Hurwitz Theorem (Theorem 2.15 in Section 2.4), the limit function of ψ_k, $\sigma_0(z) - \sigma_0(Q)$, is either equal to zero identically or never zero. It is not possible that σ_0 is equal to zero identically since $\sigma_0'(z_0) = M > 0$. Thus, for every $z \in \bar{D}(R, s)$ we have $\sigma_0(z) \neq \sigma_0(Q)$, especially $\sigma_0(R) \neq \sigma_0(Q)$. Since R, Q are arbitrary, σ_0 is univalent.

Finally, we show that σ_0 maps U onto $D(0, 1)$.

Since U is simply connected, if F is a non-zero holomorphic function on U, then we can define $\log F$ on U by

$$\log F(z) = \int_{\gamma_z} \frac{F'(\zeta)}{F(\zeta)}\, d\zeta + \log F(z_0),$$

where γ_z is a piecewise C^1 curve from z_0 to z. Since U is simply connected, the definition of $\log F(z)$ does not depend on the choice of the path γ_z. The definition of the αth power of $F(z)$ is

$$F^\alpha(z) = \exp(\alpha \log F(z)),$$

where $\alpha \in \mathbb{C}$.

If σ_0 does not map U onto $D(0, 1)$, then there exists a point $\beta \in D(0, 1)$, but $\beta \notin \sigma_0(U)$. Let

$$\varphi_\beta(\zeta) = \frac{\zeta - \beta}{1 - \bar{\beta}\zeta}.$$

Then $\mu(\zeta) = (\varphi_\beta \circ \sigma_0(\zeta))^{1/2}$ is a holomorphic function defined on U. Let $\tau = \mu(z_0)$ and

$$\varphi_\tau(\zeta) = \frac{\zeta - \tau}{1 - \bar{\tau}\zeta}, \quad \eta(\zeta) = \varphi_\tau \circ \mu(\zeta),$$

$$\nu(\zeta) = \frac{|\eta'(z_0)|}{\eta'(z_0)}\eta(\zeta).$$

Then $\nu \in \mathcal{F}$. But $\nu(z_0) = 0$ and

$$|\nu'(z_0)| = \frac{1 + |\beta|}{2|\beta|^{1/2}}M > M.$$

This contradict with the definition of σ_0. Therefore, σ_0 is onto. Such a σ_0 is just the f that is needed in the Riemann Mapping Theorem. If there is another univalent holomorphic function $g(z)$ which has the same property, then $F(z) = f(g^{-1}(z))$ is an automorphism on the unit disc and $F(0) = 0$, $F'(0) > 0$. By the Theorem 2.20 in Section 2.5, $F(z) = z$.

If U is an unbounded region, then it always can be transformed to a bounded region.

Since there are at least two points on the boundary, without loss of generosity, we can assume that they are 0 and $a\,(\neq \infty)$. Otherwise, we can always map the two boundary points to 0 and a by a fractional linear transformation.

Let $g(z)$ be a branch of $\sqrt{z-a}$ on U. Since U is simply connected, so is $g(U)$. Obviously, $g(z)$ is univalent on U. In fact, if $z_1, z_2 \in U$, $z_1 \neq z_2$, but $\sqrt{z_1 - a} = \sqrt{z_2 - a}$, then $z_1 - a = z_2 - a$ and $z_1 = z_2$. this is a contradiction. We also claim that $g(U) \bigcap (-g(U)) = \emptyset$. Suppose the contrary. Then there exists a $P \in g(U)$ with $-P \in g(U)$ and we can find $z_1, z_2 \in D$ such that $\sqrt{z_1 - a} = P$, $\sqrt{z_2 - a} = -P$. It follows that $z_1 = z_2$. Thus $P = -P$ and $P = 0$. This is a contradiction since $0 \in \partial U$.

For an arbitrary point $q \in g(U)$, since $g(U)$ is simply connected, there exists a neighborhood of $g(q)$, $U_q \subset g(U)$. Since $-U_q \not\subset g(U)$, we can choose a point b in $-U_q$ and define a function $\varphi(z) = 1/(z-b)$. This function maps $g(U)$ to a bounded simply connected region. Thus $\varphi \circ g$ maps U univalent to a bounded simply connected region and the theorem follows.

Riemann Mapping Theorem tells us that there exist univalent holomorphic functions that provides a one-to-one correspondence of the simply connected region U onto the unit disc. Does this correspondence still hold for points on the boundary? Since the boundary of a simply connected region can be very complicated, we only mention a simple case here and omit the proof.

Let U be the interior region of a Jordan curve Γ. If $w = f(z)$ is a univalent holomorphic function that maps U onto the unit disc $D(0,1)$, then $f(z)$ can be expanded to Γ and $f(z)$ is continuous on $\bar{U}$. Moreover, it maps the points of Γ one-to-one to the points of the unit circle $|w| = 1$.

4.4 Symmetry Principle

Theorem 4.5 ***(Painlevé Theorem)*** *Let U_1 and U_2 be regions with $U_1 \bigcap U_2 = \emptyset$ and $\partial U_1 \bigcap \partial U_2 = \gamma$ (Figure 8), where γ is a rectifiable curve (not including the end points). If f_1, f_2 are holomorphic on U_1, U_2 respectively and are continuous on $U_1 \bigcup \gamma, U_2 \bigcup \gamma$ and $f_1(z) = f_2(z)$ on γ, then the function*

$$f(z) = \begin{cases} f_1(z), & \text{if } z \in U-1, \\ f_1(z) = f_2(z), & \text{if } z \in \gamma, \\ f_2(z), & \text{if } z \in U-2 \end{cases}$$

is holomorphic on $U_1 \bigcup U_2 \bigcup \gamma$. The functions f_1 and f_2 are called the analytic continuations of each other across the boundary γ.

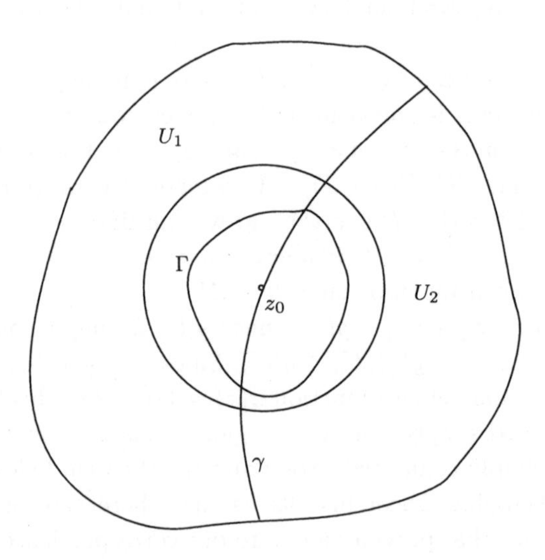

Fig. 8

Proof By the given conditions, f is continuous on $U_1 \bigcup U_2 \bigcup \gamma$ and holomorphic on U_1 and U_2. So we only need to prove that $f(z)$ is holomorphic on γ.

Let $z_0 \in \gamma$ and choose a $r > 0$ such that $D(z_0, r) \subset U\,(= U_1 \bigcup U_2 \bigcup \gamma)$. Suppose Γ is an arbitrary rectifiable simple closed curve in $D(z_0, r)$. If Γ lies inside of $U_1 \bigcup \gamma$, then by the Cauchy Integral Theorem we have

$$\int_\Gamma f(z)\,dz = \int_\Gamma f_1(z)\,dz = 0.$$

Similarly, if Γ lies inside of $U_2 \bigcup \gamma$, then

$$\int_\Gamma f(z) = dz = \int_\Gamma f_1(z)\,dz = 0.$$

If Γ lies in both U_1 and U_2, Γ_1 is the part of Γ in U_1, Γ_2 is the part of Γ in U_2 and Γ_0 is the part of γ inside Γ, then we have

$$\int_\Gamma f(z)\,dz = \int_{\Gamma_1+\Gamma_0} f_1(z)\,dz + \int_{\Gamma_2-\Gamma_0} f_2(z)\,dz = 0.$$

Thus, by the Morera Theorem (Theorem 2.9 in Section 2.3), $f(z)$ is holomorphic on $D(z_0, r)$. Especially, it is holomorphic at $z = z_0$. Since z_0 is an arbitrary point on γ, $f(z)$ is holomorphic on U.

Painlevé Theorem implies the next theorem.

Theorem 4.6 ***(Symmetry Principle)*** *Suppose a region U lies on one side of the real axis and its boundary contains a line segment s of real axis (not including the end points). If $f(z)$ is holomorphic on U, continuous on $U \bigcup s$ and the values of $f(z)$ on s are real, then there exists a function $F(z)$ which is holomorphic on $U \bigcup U' \bigcup s$ and $F(z) = f(z)$ in U, where U' is a region that is symmetric to U with respect to the real axis and $F(\bar{z}) = \overline{F(z)}$.*

Proof Define

$$F(z) = \begin{cases} f(z), & \text{if } z \in U \bigcup s, \\ \overline{f(\bar{z})}, & \text{if } z \in U' \end{cases}$$

on $U \bigcup U' \bigcup s$. This function satisfies $F(\bar{z}) = \overline{F(z)}$. We only need to show that $F(z)$ is holomorphic on $U \bigcup U' \bigcup s$. We prove $F(z)$ is holomorphic on U' first. If $z_0 \in U'$ and z is a point in a neighborhood of z_0, then

$$\frac{F(z) - F(z_0)}{z - z_0} = \frac{\overline{f(\bar{z})} - \overline{f(\bar{z_0})}}{z - z_0} = \overline{\left(\frac{f(\bar{z}) - f(\bar{z_0})}{\bar{z} - \bar{z_0}}\right)}.$$

Thus

$$\lim_{z \to z_0} \frac{F(z) - F(z_0)}{z - z_0} = \overline{f'(\bar{z_0})}.$$

Since the values of $f(z)$ are real on s, it follows that $\overline{f(x_0)} = f(x_0)$ for $x_0 \in s$. So we have

$$\lim_{\substack{z \to x_0 \\ z \in U'}} F(z) = \lim_{\substack{z \to x_0 \\ z \in U''}} \overline{f(\bar{z})} = \overline{\lim_{\substack{z \to x_0 \\ z \in U'}} f(\bar{z})} = \overline{f(x_0)} = f(x_0)$$

for $z \in U'$. Thus $F(z)$ continuous on $U' \bigcup s$. By the Painlevé Theorem (Theorem 4.5), $F(z)$ is holomorphic on $U \bigcup U' \bigcup s$.

Theorem 4.6 can also be stated in a generalized form.

Theorem 4.7 ***(Symmetry Principle)*** *Suppose U lies on one side of a straight line l, and its boundary contains a line segment s of l (not including the end points). If $f(z)$ is holomorphic on U, continuous on $U \bigcup s$ and the values of $f(z)$ on s is on L, where L is a straight line on the image plane, then there exists a function $F(z)$ that is holomorphic on $U \bigcup U' \bigcup s$ and $F(z) = f(z)$ in U, where U' is a region symmetric to U with respect to l. If z_1 and z_2 are two points which are symmetric with respect to l in $U \bigcup U' \bigcup s$, then $F(z_1)$ and $F(z_2)$ are two symmetric points with respect to L.*

Proof Transform l to the real axis by $Z = az + b$ and transform L to the real axis by $W = cw + d$. Apply the Symmetric Principle to Z and W, then transform back to z and w and the theorem follows.

The Symmetry Principle can also be generalized by changing s in Theorem 4.6 into an arc and change l in Theorem 4.7 into an arc. The analytic continuation can be obtained in a similar way.

4.5 Some Examples of Riemann Surface

According to Riemann Mapping Theorem, any two simply connected region (whose boundaries contain more than one point) are holomorphically equivalent. That is, there exist univalent holomorphic maps which map one to the other. If a holomorphic map between two regions is not univalent, then it is a multi-valued or infinite-valued function. We need the concept of Riemann surface to establish a one-to-one correspondence between the two regions of the given map. The concept of Riemann surface is a very important one in complex analysis. A Riemann surface is essentially a one dimensional complex manifold, the subject of complex analysis in a more general sense. For a more systematic treatment of Riemann surfaces the readers are referred to L. V. Ahlfors and L. Sario [2], H. Wu, I. N. Lu and Z. H. Chen [3], G. Springer [4]. We now describe Riemann surface through some examples.

By the discussion of elementary function in Section 1.5, we have already seen that $w = z^\alpha, \alpha = a + ib$ is an infinite valued function when $b \neq 0$. It is a single valued function when $b = 0$ and a is an integer. The inverse of this function is not single valued. When z varies in the angular region

$$\frac{(k-1)2\pi}{n} < \arg z < \frac{k2\pi}{n}, \quad (k = 1, 2, \cdots, n),$$

$w = z^n$ maps any of the above region to the entire w-plane except the positive part of the real axis. This map is one-to-one and holomorphic. Every image of the angular regions has a "cut" at the positive part of real axis. Thus these n angular regions correspond to n w-planes with the "cut". By the order of these n w-planes $k = 1, 2, \cdots, n$, we attach the lower edge of the cut of each w-plane to the upper edge of the cut of the next w-plane. The lower edge of the cut of the nth plane is attached to the upper edge of the cut of the first plane (It seems impossible to do so unless these planes are self-intersected. But here we mean to identify the lower edge of the n-th plane with the upper edge of the first plane). Now, a Riemann surface is constructed. We call the w-planes the *sheets* or *branches* of the Riemann surface. It is clear that when a point z varies on the z-plane, the corresponding point w varies on the Riemann surface. And all the points on z-plane one-to-one corresponds to the points on the Riemann surface.

We can also perform a cut along any ray from 0 to ∞ instead of along the positive real axis. The Riemann surface we get this way is identical to the one constructed above. But it is necessary to mention which cut is taken before the construction of the Riemann surface.

Notice that the point $w = 0$ is special. It connects all the branches. A curve will circle around $w = 0$ n times before it becomes a closed curve. Such a point is called a *branch point*. If $w = \infty$ is included, then it is also a branch point. In general, it is not necessary for a branch point to connect all branches. If it connects h branches, then it is said to be of order $h - 1$.

Similarly, we can discuss the Riemann surface of $w = e^z$. This function maps the band region $2\pi(k-1) < y < 2\pi k$, where $z = x + iy$, to a branch of w-plane with the cut along the positive real axis. There are infinitely many branches attached together. Since e^z never vanishes, $w = 0$ is not in the Riemann surface.

Conversely, the function $w = z^{1/n}$, where $n > 1$ is a positive integer, maps the Riemann surface of n branches to the w-plane and the map is one-to-one. Similarly, $w = \log z$ also maps a Riemann surface with infinitely many branches to the w-plane and the map is one-to-one.

Thus, when we talk about the correspondence between Riemann surface and the complex plane, it is usually necessary to point out which branch of Riemann surface is used.

4.6 Schwarz-Christoffel Formula

The Riemann Mapping Theorem is an existence theorem. It is not quite easy to write down the concrete map. In Section 4.1, we looked at some basic examples. In this section, we give a formula that maps the upper half plane to a polygon. It is called the *Schwarz-Christoffel Formula.*

Suppose that $-\infty < a_1 < a_2 < \cdots < a_n < \infty$ are n real numbers, $a_0 = -\infty$ and $a_{n+1} = +\infty$. If $\alpha_1, \alpha_2, \cdots, \alpha_n$ are n positive real numbers satisfying $\alpha_1 + \alpha_2 + \cdots + \alpha_n + 1 < n$, and

$$\beta(t) = (t - a_1)^{\alpha_1 - 1} \cdots (t - a_n)^{\alpha_n - 1}, \tag{4.5}$$

then

$$\int_{-\infty}^{\infty} |\beta(t)|\, dt < \infty.$$

For $t < a_k$, choose such a branch of

$$(t - a_k)^{\alpha_k - 1} = \exp((\alpha_k - 1)\log(t - a_k))$$

that the argument is $\pi(\alpha_k - 1)$. Thus, when $t < a_1$, we have

$$\arg \beta(t) = \pi[(\alpha_1 + \cdots + \alpha_n) - n].$$

When t is on the line segment of (a_{k-1}, a_k), we have

$$\arg \beta(t) = \pi[(\alpha_k + \cdots + \alpha_n) - (n - k + 1)], \quad 2 \le k \le n.$$

When t is on (a_n, ∞), we have

$$\arg \beta(t) = 0.$$

Therefore, we have defined $n+2$ complex numbers

$$w_k = c\int_0^{a_k} (t-a_1)^{\alpha_1 - 1}\cdots(t-a_n)^{\alpha_n - 1}\,dt, \quad 0 \le k \le n+1, \tag{4.6}$$

where c is a positive number. Define a function

$$f(z) = c\int_0^z \beta(t)\,dt \tag{4.7}$$

on the upper half plane $\bar{H} = \{z \in \mathbb{C} |\ \mathrm{Im}\, z \ge 0\}$. This function is obviously holomorphic on $H = \{z \in \mathbb{C} |\ \mathrm{Im}\, z > 0\}$. On the real axis, we have that

$$\begin{aligned} f(x) &= w_{k-1} + c\int_{a_k - 1}^{x} \beta(t)\,dt \\ &= w_{k-1} + ce^{i[(\alpha_k - 1)\pi + \cdots + (\alpha_n - 1)\pi]}\int_{a_k - 1}^{x} |\beta(t)|\,dt \end{aligned}$$

when $x \in (a_{k-1}, a_k)\,(1 \le k \le n+1)$. Thus, $f(x) - w_{k-1}$ has the same argument $[(\alpha_k - 1)\pi + \cdots + (\alpha_n - 1)\pi]$ on the interval (a_{k-1}, a_k), and its absolute value increases from 0 to

$$l_k = c\int_{a_{k-1}}^{a_k} |\beta(t)|\,dt. \tag{4.8}$$

Therefore, when x varies in the interval $[a_{k-1}, a_k]$, f varies in the interval $\Delta_{k-1} = [w_{k-1}, w_k]$. The argument of Δ_{k-1} is $(\alpha_k - 1)\pi + \cdots + (\alpha_n - 1)\pi$ and the length is l_k.

To prove $w_0 = w_{n+1}$, we only need to show that for any $\varepsilon > 0$, there exists an $R > 0$ such that $|w_0 - f(z)| < \varepsilon$ when $z \in \bar{H}$ and $|z| \ge R$ because this is equivalent to $\lim_{z\to\infty} f(z) = w_{n+1} = w_0$. Indeed, since

$$\int_{-\infty}^{\infty} |\beta(t)|\,dt < \infty,$$

for any $\varepsilon > 0$, there exists an $R_1 > 0$ such that

$$\int_{-\infty}^{-R_1} |\beta(t)|\,dt \le \frac{\varepsilon}{2}.$$

Thus

$$|w_0 - c\int_0^x \beta(t)\,dt| \le \int_{-\infty}^{-R_1} |\beta(t)|\,dt \le \frac{\varepsilon}{2}$$

when $-\infty < x < -R$. Let $R_1 \geq \max\{|a_1|, \cdots, |a_n|\}$ and $z_0 = \rho_0 e^{i\theta_0}$ where $\rho_0 \geq R_1$ and $0 \leq \theta_0 \leq \pi$. Then

$$|f(z_0) - f(\rho_0)| = c\left|\int_0^{\theta_0} (\rho_0 e^{i\theta} - a_1)^{\alpha_1 - 1} \cdots (\rho_0 e^{i\theta} - a_n)^{\alpha_n - 1} \rho_0 e^{i\theta}\, d\theta\right|$$
$$\leq c\rho_0(\rho_0 - R_1)^{\alpha_1 + \cdots + \alpha_n - n}.$$

Since $\alpha_1 + \cdots + \alpha_n < n - 1$, $c\rho_0(\rho_0 - R_1)^{\alpha_1+\cdots+\alpha_n-n}$ tends to zero as $\rho_0 \to \infty$. Thus, there exists an $R \geq R_1$ such that $|f(z_0) - f(\rho_0)| \leq \varepsilon/2$ when $\rho_0 = |z_0| \geq R$. Hence, $|f(z) - w_0| \leq \varepsilon$ when $|z| \geq R$ and $\operatorname{Im} z \geq 0$. This proves that $w_0 = w_{n+1}$. Therefore, f maps $R\bigcup\{\infty\}$ to the $(n+1)$-polygon with sides $\Delta_0, \Delta_1, \cdots, \Delta_n$ and vertices $w_0, w_1, \cdots, w_n, w_{n+1} = w_0$.

If $0 < \alpha_k < 2$, then we can show that the interior angle of this polygon at the vertex w_k is $\alpha_k \pi$. In fact, since $\beta(t)$ can be written as $\beta_k(t)(t - a_k)^{\alpha_k - 1}$ where $\beta_k(t) = \prod_{j \neq k}(t - a_j)^{\alpha_j}$, $\beta(t)$ is holomorphic in a neighborhood V of a_k and it can be expanded to the Taylor series

$$\beta_k(t) = a_{0,k} + a_{1,k}(t - a_k) + \cdots, \qquad a_{0,k} \neq 0.$$

Thus, when $z \in V \bigcap \bar{H}$, we have

$$f(z) = w_k + c\int_{a_k}^{z} (a_{0,k} + a_{1,k}(t - a_k) + \cdots)(t - a_k)^{\alpha_k - 1}\, dt$$
$$= w_k + c\frac{a_{0,k}}{\alpha_k}(z - a_k)^{\alpha_k}\left(1 + \frac{\alpha_k}{\alpha_k + 1} \cdot \frac{a_{1,k}}{a_{0,k}}(z - a_k) + \cdots\right).$$

As z approaches to a_k along a straight line with inclination θ $(0 \leq \theta \leq \pi)$, $f(z)$ approaches to w_k along a curve. The inclination of tangent line at the point w_k is $\arg(a_{0,k}) + \alpha_k\theta$ since $c > 0$ and $\alpha_k > 0$. Draw a half disc in $\bar{H}$ with center a_k that is contained completely in $V \bigcap \bar{H}$. Let z travels along the boundary of this half disc from $\theta = 0$ to $\theta = \pi$. Then $f(z)$ travels along a Jordan curve from the point of Δ_{k-1} to the point of Δ_k and the argument of $f(z)$ also changes from $\arg(a_{0,k})$ to $\arg(a_{0,k}) + \alpha_k\pi$. Thus, the interior angle at the vertex w_k is $\alpha_k\pi$. the interior angle at the vertex $w_0 = w_{n+1}$ is $((n-1) - (\alpha_1 + \cdots + \alpha_n))\pi > 0$ since the sum of the interior angle of a polygon with $(n+1)$ sides is $(n+1)\pi$. Especially, the interior angle at the vertex w_0 is π when $\alpha_1 + \cdots + \alpha_n = n - 2$. This is just a polygon with n sides.

Equation (4.7) is called the *Schwarz-Christoffel Formula* where $\beta(t)$ is defined by (4.5) and $\alpha_1 + \cdots + \alpha_n < n - 1$. The map (4.7) sends the upper half plane to a polygon with $(n+1)$ sides whose vertices are

$w_0, w_1, \cdots, w_n, w_{n+1} = w_0$, which are defined by (4.6). The sides of this polygon are $[w_{k-1}, w_k]$ $(k = 1, \cdots, n+1)$ and their length are l_k which is defined by (4.8). If $0 < \alpha_j < 2$ $(j = 1, \cdots, n)$, then the interior angle at the vertex w_k is $\alpha_k \pi$ $(k = 1, \cdots, n)$. The interior angle at $w_0 = w_{n+1}$ is $[(n-1) - (\alpha_1 + \cdots + \alpha_n)]\pi$.

The equation (4.7) can be written in a generalized form

$$f(z) = c \int_0^z \beta(t)\, dt + c'. \tag{4.9}$$

If we write (4.6) in a generalized form

$$w_k = c \int_0^{a_k} (t - a_1)^{\alpha_1 - 1} \cdots (t - a_n)^{\alpha_n - 1}\, dt + c',\ 0 \leq k \leq n+1,$$

where c and c' are complex numbers, then (4.9) is also Schwarz-Christoffel Formula.

Exercise IV

1. Verify the examples in Section 4.1.

2. Prove that the cross ratio is an invariant of fractional linear transformations.

3. Show that if the entire function $f(z) = \sum_{n=0}^{\infty} c_n z^n$ assumes real values on the real axis, then c_n $(n = 0, 1, \cdots)$ are real numbers.

4. Find the Riemann surface of $w = z + \sqrt{z^2 - 1}$.

5. Show that any circle can be represented as

$$\left| \frac{z - z_1}{z_1 - z_2} \right| = k\ (k > 0)$$

where z_1 and z_2 are symmetric with respect to the circle. The center and the radius of this circle are

$$a = \frac{z_1 - k^2 z_2}{1 - k}, \qquad R = \frac{k|z_1 - z_2|}{|1 - k^2|}$$

respectively.

6. Show that the cross ratio (z_1, z_2, z_3, z_4) is a real number if and only if the four points z_1, z_2, z_3 and z_4 are on the same circle.

7. (*Carathéodory Inequality*) Using Schwarz Lemma and the linear transformation, show that if a function $f(z)$ is holomorphic in $|z| < R$, continuous on $|z| \le R$, $M(r)$ and $A(r)$ are the maximal values of $|f(z)|$ and $\operatorname{Re} f(z)$ on $|z| = r$ respectively, then

$$M(r) \le \frac{2r}{R-r} A(R) + \frac{R+r}{R-r} |f(0)|$$

when $0 < r < R$.

8. Suppose that the function $f(z) = \sum_{n=0}^{\infty} a_n z^n$ is holomorphic in $|z| < 1$ and $|f(z)| \le M$. Show that $M|a_1| \le M^2 - |a_0|^2$.

9. Find the following conformal mappings:

(i) From the band region $0 < \operatorname{Im} z < \pi$ on z-plane to $|w| < 1$.

(ii) From the half disc $|z| < 1$, $\operatorname{Im} z > 0$ on the z-plane to the upper half plane.

(iii) From the intersection of $|z| < 1$ and $|z - 1| < 1$ on z-plane to $|w| < 1$.

(iv) From the fan shaped region $\{z |\ 0 < \arg z < \alpha\, (< 2\pi),\ |z| < 1\}$ on z-plane to $|w| < 1$.

(v) From $|z| < 1$ on z-plane to the belt region $0 < v < 1$, where $w = u + iv$, and it maps $-1, 1, i$ to $\infty \infty, i$.

(vi) From $\mathbb{C} \setminus (-\infty, -(1/4)]$ on z-plane to $|w| < 1$.

10. Find the following conformal mappings which send the upper half plane to the polygon P by using the Schwarz-Christoffel Formula.

(i) P is a triangle with vertices w_1, w_2, w_3 and the corresponding angles $\alpha\pi, \beta\pi, \gamma\pi$, where $\alpha + \beta + \gamma = \pi$.

(ii) P is an equilateral triangle with vertices $w_1 = 0$, $w_2 = a$, $w_3 = ((1 + \sqrt{3}i)/2)a$, where $a > 0$.

(iii) P is an isosceles triangle with vertices $w_1 = 0$, $w_2 = a$, $w_3 = a(1+i)$, where $a > 0$.

(iv) P is a rectangle with vertices $w_1 = -k_1$, $w_2 = k_1$, $w_3 = k_1 + ik_2$, $w_4 = -k_1 + ik_2$, where $k_1 > 0$, $k_2 > 0$.

(v) P is a rhombus with vertices $0, a, a(1 + e^{i\alpha\pi}), ae^{i\alpha\pi}$, where $0 < \alpha < \pi/2$ and $a > 0$.

(vi) P is the inside of a pentagon with the center at the origin and $w = 1$ is one of the vertices of P.

11. In the Riemann Mapping Theorem, suppose that z_0 is a real number and U is symmetric about the real axis. Show that f satisfies the symmetric property $f(\bar{z}) = \overline{f(z)}$ by using the uniqueness of the theorem.

12. Suppose $\mathcal{F}$ is a collection of all holomorphic functions defined on a region Ω with values on the right half plane. If there exists a point $z_0 \in \Omega$ such that $f(z_0) = g(z_0)$ for all $f, g \in \mathcal{F}$, then $\mathcal{F}$ is a normal family.

13. Let $\mathcal{F}$ be a collection of all holomorphic functions defined on a region Ω with values in $U_0 = \mathbb{C} \setminus \{x + i0,\ 0 \leq x \leq 1\}$. If there exists a point $z_0 \in \Omega$ such that $f(z_0) = g(z_0)$ for all $f, g \in \mathcal{F}$, then $\mathcal{F}$ is a normal family.

Appendix II Riemann Surface

In Section 4.5 of this chapter, we described the Riemann surface by some examples. We now define Riemann surface formally.

Let X be a set, $\mathcal{F}$ be a collection of subsets of X. If $\mathcal{F}$ has the following properties: (1) X and $\emptyset$ are in $\mathcal{F}$; (2) The union of the elements of any subcollection of $\mathcal{F}$ is in $\mathcal{F}$; (3) The intersection of the elements of any finite subcollection of $\mathcal{F}$ is in $\mathcal{F}$, then $\mathcal{F}$ is called a *topology* on X. The elements of $\mathcal{F}$ are called the *open sets.* A set X for which a topology $\mathcal{F}$ has been specified is called a *topological space* and is denoted by $(X, \mathcal{F})$.

In brief, a topological space is a set in which open sets are defined.

Let U be a subset of X and $x \in X$. If there exists an open subset G of X such that $x \in G \subset U$, then U is called a *neighborhood* of x.

A topological space is called a *Hausdorff space* if any two distinct points have distinct neighborhoods.

Let X, Y be topological spaces, f be a map from X to Y which is one-to-one and onto. Assume that both f and f^{-1} are continuous. Then f is called a *homeomorphism* from X to Y.

A *surface* is a Hausdorff space which is locally homeomorphic to $\mathbb{R}^2$. Here, locally homeomorphic means that for each point of the space there exists a neighborhood that is homeomorphic to an open subset of $\mathbb{R}^2$.

A Riemann surface is a connected Hausdorff topological space C together with an open covering $\{U_\alpha\}$ and a collection of maps $f_\alpha :\ U_\alpha \to C$ that satisfy the following properties:

(a) $f :\ U_\alpha \to C$ is a homeomorphic map from U_α to C;

(b) if $U_\alpha \bigcap U_\beta \neq \emptyset$, then the function

$$f_\beta \circ f_\alpha^{-1} :\ f_\alpha(U_\alpha \bigcap U_\beta) \to f_\beta(U_\alpha \bigcap U_\beta)$$

is biholomorphic (both the function and its inverse are holomorphic). We call (U_α, f_α) a *holomorphic local coordinate system* and $\{(U_\alpha, f_\alpha)\}$ a *holomorphic coordinate covering.*

In brief, Riemann surface is a locally Euclidean Hausdorff space with a complex structure.

This definition of Riemann surface is also the definition of one dimensional complex manifold. Higher dimensional complex manifolds can be defined in a similar way.

Let's look at an example of Riemann surface.

Example 4.4 The extended complex plane $\mathbb{C}^* = \mathbb{C} \bigcup \{\infty\}$ is a Riemann surface.

It is easy to verify that $\mathbb{C}^*$ is a connected Hausdorff topological space. Choose $\{U_0, U_1\}$ as an open covering of $\mathbb{C}^*$ where

$$U_0 = \mathbb{C}, \quad U_1 = \mathbb{C}^* \setminus \{0\}.$$

Let $f_0(z) = z$ and

$$f_1(z) = \begin{cases} 0, & \text{if } z = \infty, \\ \frac{1}{z} & \text{if } z \neq \infty. \end{cases}$$

Then obviously both $f_1 \circ f_0^{-1}$ and $f_0 \circ f_1^{-1}$ are biholomorphic on $\mathbb{C} \setminus \{0\}$. Hence $\mathbb{C}^*$ is a Riemann surface. It is easy to verify that the Riemann sphere is also a Riemann surface.

The reader is referred to related books for further discussions about the definition of derivative, integral, meromorphic functions, zeros, poles and the classifications of Riemann surfaces.

Chapter 5

Differential Geometry and Picard Theorem

5.1 Metric and Curvature

The Picard Theorem is one of the important classical theorems in complex analysis and especially in value distribution theory. The original proof of this theorem is complicated. In this chapter, we will give a straight forward proof by using the method of differential geometry. First we need some basic knowledge of complex differential geometry.

Let Ω be a region in $\mathbb{C}$. Define a non-negative C^2 function ρ on Ω such that $ds_\rho^2 = \rho^2|dz|^2$ and call it a *metric*. From this, we can define the distance function d. The distance between two points z_1 and z_2 in Ω is

$$d(z_1, z_2) = \inf \int_\gamma \rho(z)|dz|. \tag{5.1}$$

The infimum is taken over all the curves γ entirely contained in Ω that connect z_1 and z_2.

The *curvature* with respect to the metric ρ is defined by

$$K(z, \rho) = -\frac{\Delta \log \rho(z)}{\rho^2(z)}, \tag{5.2}$$

where Δ is the Laplace operator

$$\begin{aligned}\Delta &= \frac{\partial^2}{\partial x^2} + \frac{\partial^2}{\partial y^2} = 4\frac{\partial}{\partial z}\frac{\partial}{\partial \bar{z}} \\ &= 4\frac{\partial}{\partial \bar{z}}\frac{\partial}{\partial z} = \frac{\partial^2}{\partial r^2} + \frac{1}{r}\frac{\partial}{\partial r} + \frac{1}{r^2}\frac{\partial^2}{\partial \theta^2},\end{aligned}$$

where $z = x + iy = re^{i\theta}$.

The curvature under this definition is the same as the Gauss curvature in differential geometry (cf. the Appendix of this chapter).

The following three metrics are often used in complex differential geometry.

1. Euclidean Metric

If $\Omega = \mathbb{C}$, we define the *Euclidean metric* (or the *parabolic metric*) to be $\rho(z) = 1$, i.e., $ds^2 = |dz|^2$, for all $z \in \mathbb{C}$. The distance between two points z_1 and z_2 is

$$d(z_1, z_2) = \inf \int_\gamma |dz| = |z_1 - z_2| = l,$$

where l is the length of the line segment that joins z_1 and z_2. This is called the *Euclidean distance.*

The group formed by transformations of the form $w = e^{i\theta}z + a$ where θ is a real number and a is a complex number is called the *group of Euclidean motions* or the *group of rigid motions*. In this group, each element is the composition of a rotation $w = e^{i\theta}z$ and a translation $w = z + a$. This is a subgroup of $\mathrm{Aut}(\mathbb{C})$. Obviously, the Euclidean metric is an invariant of the group of Euclidean motions. By definition (5.2), $K(z, \rho) = 0$ for any $z \in \mathbb{C}$. Therefore, this metric is also called the parabolic metric.

2. Poincaré Metric

Let Ω be the unit disc $D(0,1) = \{z| \, |z| < 1\}$. We define the *Poincaré metric* (or the *hyperbolic metric*) to be

$$\rho = \lambda(z) = \frac{2}{1 - |z|^2}, \; ds_\lambda^2 = \frac{4|dz|^2}{(1 - |z|^2)^2}$$

on $D(0,1)$. It was shown in Section 2.5 that the group of holomorphic automorphisms of $D(0,1)$, $\mathrm{Aut}(D(0,1))$, is formed by transformations of the form

$$w = e^{i\theta}\frac{z - a}{1 - \bar{a}z}$$

where θ is a real number and $a \in D(0,1)$. It is formed by rotations and Möbius transformations. It was also shown in Section 2.5 that the Poincaré metric is an invariant of $\mathrm{Aut}(D(0,1))$.

Now, let us calculate the Poincaré distance between two points z_1 and z_2 in $D(0,1)$.

First, consider the distance between two points $z_1 = 0$ and $z_2 = R + i0 \, (R < 1)$ in $D(0,1)$. The curve which connects these two points can be written as

$$z(t) = u(t) + iw(t), \; 0 \le t \le 1,$$

$$w(0) = u(0) = w(1) = 0, \ u(1) = R,$$

where $u^2(t) + w^2(t) < 1$ and u, w are real valued C^1 functions of t. Thus, we have

$$\begin{aligned}\int_\gamma ds &= \int_\gamma \frac{2|dz|}{1-|z|^2} = 2\int_0^1 \frac{((u'(t))^2 + (w'(t))^2)^{1/2}\,dt}{1-(u(t))^2-(w(t))^2} \\ &\geq \int_0^1 \frac{2|u'(t)|\,dt}{1-(u(t))^2} \geq \left|\int_0^R \frac{2du}{1-u^2}\right| = \log\frac{1+R}{1-R}.\end{aligned}$$

The equality holds if and only if $w(t) = 0$, $0 \leq t \leq 1$. Therefore, we have

$$d(0, R+i0) = \inf_\gamma \int_\gamma \frac{|dz|}{1-|z|^2} = \log\frac{1+R}{1-R},$$

where the integral attains its infimum on the line segment γ that connects 0 and $R + i0$.

Since $w = e^{i\theta}z$ is an element of $\mathrm{Aut}(D(0,1))$, the Poincaré distance between any two points in $D(0,1)$ does not change under the transformation $w = e^{i\theta}z$. Thus, we have

$$d(0, e^{i\theta}R) = \log\frac{1+R}{1-R}$$

holds for any real number θ.

Let z_1, z_2 be two arbitrary points in $D(0,1)$. Then

$$\varphi(z) = \frac{z-z_1}{1-\bar{z}_1 z}$$

is an element of $\mathrm{Aut}(D(0,1))$ which maps z_1 to 0 and z_2 to $(z_2-z_1)/(1-\bar{z}_1 z_2)$. Thus,

$$d(z_1, z_2) = d\left(0, \frac{z_2-z_1}{1-\bar{z}_1 z_2}\right) = \log\frac{1+\left|\dfrac{z_2-z_1}{1-\bar{z}_1 z_2}\right|}{1-\left|\dfrac{z_2-z_1}{1-\bar{z}_1 z_2}\right|}. \tag{5.3}$$

This is the Poincaré distance (or the hyperbolic distance) between any two points z_1, z_2 of $D(0,1)$. In this case, the curve γ on which

$$d(z_1, z_2) = \inf_\gamma \int_\gamma \frac{|dz|}{1-|z|^2}$$

attains its infimum is

$$z = \frac{z_1 + \dfrac{z_2 - z_1}{1 - \bar{z_1} z_2} t}{1 + \bar{z_1} \dfrac{z_2 - z_1}{1 - \bar{z_1} z_2} t}, \ 0 \leq t \leq 1.$$

That is

$$z = \frac{(1-t) z_1 + (t - z_1 \bar{z_1}) z_2}{1 - t z_1 \bar{z_1} - (1-t) \bar{z_1} z_2}, \ 0 \leq t \leq 1.$$

From (5.3), we can see that $d(z_1, z_2) \to 0$ as $z_2 \to z_1$ and $d(z_1, z_2) \to \infty$ as z_1 or z_2 approach to the boundary of $D(0, 1)$.

Recall Schwarz-Pick Lemma (Theorem 2.21 in Section 2.5): If $w = f(z)$ is a holomorphic function on $D(0, 1)$ which maps $D(0, 1)$ into $D(0, 1)$ and $w_1 = f(z_1)$, $w_2 = f(z_2)$, then

$$\left| \frac{w_1 - w_2}{1 - \bar{w_1} w_2} \right| \leq \left| \frac{z_1 - z_2}{1 - \bar{z_1} z_2} \right|. \tag{5.4}$$

The equality holds if and only if $f(z) \in \mathrm{Aut}(D(0, 1))$.

By (5.3), we can rewrite (5.4) as

$$d(w_1, w_2) \leq d(z_1, z_2).$$

Therefore, we get the geometric meaning of Schwarz-Pick Lemma: If $w = f(z)$ is a holomorphic function which maps $D(0, 1)$ into $D(0, 1)$, then the Poincaré distance between two arbitrary points in $D(0, 1)$ does not increase under f. The distance does not change if and only if $f(z) \in \mathrm{Aut}(D(0, 1))$.

Since

$$\Delta = 4 \frac{\partial}{\partial z} \frac{\partial}{\partial \bar{z}},$$

we have

$$-\Delta \log \lambda(z) = \Delta \log(1 - |z|^2) = \frac{-4}{(1 - |z|^2)^2}.$$

Thus, the curvature $K(z, \lambda)$ of the hyperbolic metric $\lambda(z)$ is equal to -1 for all $z \in D(0, 1)$. Therefore, this metric is also called the hyperbolic metric.

3. Spherical Metric

Let Ω be $\mathbb{C}^*$. We define the *spherical metric* (or the *elliptic metric*) to be

$$\rho = \sigma(z) = \frac{2}{1 + |z|^2}, \ ds_\sigma^2 = \frac{4 |dz|^2}{(1 + |z|^2)^2}.$$

The stereographic projection, which was introduced in Section 1.2, constructed a one-to-one correspondence between the points of Riemann sphere S^2 and $\mathbb{C}^*$. If $z \in \mathbb{C}^*$, then the coordinate of the corresponding point on S^2 is

$$(x_1, x_2, x_3) = \left(\frac{z+\bar{z}}{|z|^2+1}, \frac{z-\bar{z}}{i(|z|^2+1)}, \frac{|z|^2-1}{|z|^2+1} \right). \tag{5.5}$$

If $P = (x_1, x_2, x_3)$ and $P' = (x_1', x_2', x_3')$ are two points on S^2, then the shortest distance between the two points on S^2 is the length of the arc $\overset{\frown}{PP'}$ on the great circle through P and P'. The length of the arc is

$$2 \arctan \sqrt{\frac{1 - x_1x_1' - x_2x_2' - x_3x_3'}{1 + x_1x_1' + x_2x_2' + x_3x_3'}}.$$

By (5.5), this is equal to

$$2 \arctan \left| \frac{z - z'}{1 + z\bar{z}'} \right|.$$

If we consider this distance as a metric in $\mathbb{C}^*$, then the corresponding distance between z and z' is

$$d(z, z') = 2 \arctan \left| \frac{z - z'}{1 + z\bar{z}'} \right|.$$

Obviously, the corresponding metric is

$$ds^2 = \frac{4|dz|^2}{(1+|z|^2)^2}.$$

This can be obtained by taking the derivative on d,

$$ds^2 = \sigma^2(z)|dz|^2, \quad \sigma(z) = \frac{2}{1+|z|^2}.$$

Thus the metric $\sigma(z)$ has a clear geometric meaning. The distance between two points in $\mathbb{C}^*$ under the metric $\sigma(z)$ is equal to the shortest distance between the two corresponding points on S^2, i.e., their spherical distance. In other words, if we use the spherical distance of two corresponding points on S^2 as the distance of two points on $\mathbb{C}^*$, then we have

$$d(z_1, z_2) = \inf_{\gamma} \int_{\gamma} \sigma(z)|dz|,$$

where γ is an arbitrary curve joining z_1 and z_2.

Let the corresponding points of z_1 and z_2 on S^2 be P_1 and P_2 respectively. Connect P_1 and P_2 by the arc $\widehat{P_1P_2}$ of the great circle. If γ_0 is the spherical projection of $\widehat{P_1P_2}$ on $\mathbb{C}^*$, then the above integral attains its infimum on the curve γ_0. Therefore, $\sigma(z)$ is called the spherical metric.

A simple calculation shows that the curvature at every point $z \in \mathbb{C}^*$ with respect to the spherical metric $\sigma(z)$ is $+1$. Therefore this metric is also called the elliptic metric.

Recall the Uniformization Theorem in Section 3.3: Any simply connected Riemann surface is one-to-one holomorphic equivalent to one of these three regions: $\mathbb{C}$, $D(0,1)$ and $\mathbb{C}^*$. That is why we discuss the geometry on these three regions.

A important property of the curvature defined by (5.2) is that it is a invariant under holomorphic maps.

Let Ω_1 and Ω_2 be two regions of $\mathbb{C}$, f be a holomorphic function on Ω_1 that maps Ω_1 to Ω_2. If ρ is a metric on Ω_2 and $f' \neq 0$, then

$$f^*\rho = (\rho \circ f)|f'| \tag{5.6}$$

defines a metric on Ω_1. This metric is referred to as a pullback of ρ by $f(z)$. Now, we show that

$$K(z, f^*\rho) = K(f(z), \rho). \tag{5.7}$$

Since $\Delta \log |f'(z)| = 0$ and

$$\begin{aligned}
\Delta \log(\rho \circ f(z)) &= 4\frac{\partial}{\partial \bar{z}}\frac{\partial}{\partial z}\log(\rho \circ f(z)) \\
&= 4\overline{\left(\frac{\partial f}{\partial z}\right)}\left(\frac{\partial f}{\partial z}\right)\left(\frac{\partial}{\partial f}\frac{\partial}{\partial \bar{f}}\log(\rho \circ f)\right) \\
&= |f'(z)|^2(\Delta_f \log \rho) \circ f(z),
\end{aligned}$$

we have

$$\begin{aligned}
K(z, f^*\rho) &= -\frac{|f'(z)|^2(\Delta_f \log \rho) \circ f(z)}{(\rho \circ f(z))^2|f'(z)|^2} \\
&= -\frac{(\Delta_f \log \rho) \circ f(z)}{(\rho \circ f(z))^2} \\
&= K(f(z), \rho).
\end{aligned}$$

5.2 Ahlfors-Schwarz Lemma

Theorem 2.19 in Section 2.5 gives an analytic form of the classical Schwarz Lemma, and Theorem 2.21 in Section 2.5, Schwarz-Pick Lemma, gives a generalization of Schwarz Lemma. In the last section, we discussed the geometric meaning of Schwarz-Pick Lemma for Poincaré metric. In this section, we will give another form of generalization of Schwarz Lemma by using curvatures, which we refer to as Ahlfors-Schwarz Lemma. This lemma was established by Ahlfors in 1938 (L. V. Ahlfors [2]). It started the study of complex analysis from the viewpoint of differential geometry.

Theorem 5.1 ***(Ahlfors-Schwarz Lemma)*** *Let $f(z)$ be a holomorphic function on $D(0,1)$ which maps $D(0,1)$ to U. If a metric ρ is defined on U, $ds_\rho^2 = \rho^2(z)|dz|^2$, and its curvature is less than or equal to -1 at any point of U, then*

$$f^*\rho(z) \leq \lambda(z) \tag{5.8}$$

where $\lambda(z) = 2/(1-|z|^2)$, that is

$$ds_\rho^2 \leq ds_\lambda^2. \tag{5.9}$$

In other words, the metric does not increase under the map f.

Proof For any fixed $r \in (0,1)$, define the metric

$$\lambda_r(z) = \frac{2r}{r^2 - z^2}.$$

on $D(0,r)$. It is easy to see that the curvature is -1 for $z \in D(0,r)$. Let

$$v(z) = \frac{f^*\rho(z)}{\lambda_r}.$$

Then v is non-negative and continuous on $D(0,r)$. By (5.6),

$$f^*\rho(z) = \rho(f(z))|f'(z)|$$

is bounded on $\overline{D(0,r)}$ and $1/\lambda_r \to 0$ as $|z| \to 0$. Thus, we have $v \to 0$ as $|z| \to 0$. Therefore, v can only reach its maximal value M at a point τ in $D(0,r)$. If we can show that $M \leq 1$, then $v \leq 1$ on $D(0,r)$ and (5.8) will follow by letting $r \to 1-0$.

If $f^*\rho(\tau) = 0$, then $v \equiv 0$ and there is nothing to prove. Without loss of generosity, assume that $f^*\rho(\tau) > 0$. Then $K(\tau, f^*\rho)$ is well defined. Thus, by the assumption, $K(\tau, f^*\rho) \leq -1$. Since $\log v$ attains its maximal value at τ, we have

$$\begin{aligned}0 &\geq \Delta \log v(\tau) = \Delta \log f^*\rho(\tau) - \Delta \log \lambda_r(\tau)\\ &= -K(\tau, f^*\rho)\cdot(f^*\rho(\tau))^2 + K(\tau, \lambda_r)(\lambda_r(\tau))^2\\ &\geq (f^*\rho(\tau))^2 - (\lambda_r(\tau))^2.\end{aligned}$$

Thus

$$\frac{f^*\rho(\tau)}{\lambda_r(\tau)} \leq 1.$$

Therefore, $M \leq 1$ and the lemma follows.

If $U \subseteq D(0,1)$ in Ahlfors-Schwarz Lemma, then we can get Schwarz-Pick Lemma by choose $\rho = \lambda$. Thus Ahlfors-Schwarz Lemma is a generalization of Schwarz-Pick Lemma.

We can also write a more generalized form of Ahlfors-Schwarz Lemma.

On $D(0,\alpha), (\alpha > 0)$, define a metric

$$\lambda_\alpha^A(z) = \frac{2\alpha}{\sqrt{A}(\alpha^2 - |z|^2)}, \tag{5.10}$$

where $A > 0$. The curvature of this metric is $-A$ at any point of $D(0,\alpha)$.

Theorem 5.2 ***(Generalized Ahlfors-Schwarz Lemma)*** *Let $f(z)$ be a holomorphic function on $D(0,\alpha)$ that maps $D(0,\alpha)$ to U. Define a metric ρ on U, $ds_\rho^2 = \rho^2(z)|dz|^2$, such that the curvature is less than or equal to $-B$ at every point in U. Then*

$$f^*\rho(z) \leq \frac{\sqrt{A}}{\sqrt{B}}\lambda_\alpha^A(z) \tag{5.11}$$

holds for every $z \in D(0,\alpha)$, where B is a positive number.

We shall omit the proof here since it is the same as the proof of Theorem 5.1.

Ahlfors-Schwarz Lemma is also one of the comparison theorems in Differential Geometry. A number of important results such as the generalized Liouville Theorem follows from Ahlfors-Schwarz Lemma.

5.3 The Generalization of Liouville Theorem and Value Distribution

Recall the Liouville Theorem (Theorem 2.10 in Section 2.3): Any bounded entire function must be a constant function. Now, by applying Ahlfors-Schwarz Lemma, we can interpret and generalize the Liouville Theorem using curvatures.

Theorem 5.3 ***(Generalized Liouville Theorem)*** *Suppose that a entire function f maps $\mathbb{C}$ to U. If we can define a metric $\rho(z)$ on U such that for any $z \in U$, the curvature $K(z,\rho)$ satisfies $K(z,\rho) \leq -B < 0$, where B is a positive number, then f must be a constant function.*

Proof For any $\alpha > 0$, f maps $D(0,\alpha)$ into U. By the assumption, we can define a metric ρ such that the curvature $K(z,\rho) \leq -B < 0$. Thus, by Theorem 5.2,

$$f^*\rho(z) \leq \frac{\sqrt{A}}{\sqrt{B}}\lambda_\alpha^A(z).$$

Since $\lambda_\alpha^A(z) \to 0$ as $\alpha \to \infty$ by (5.10), we have $f^*\rho(z) \leq 0$. Therefore, $f^*\rho(z) = 0$. Hence f must be a constant function since it is holomorphic. The theorem is proved.

Theorem 5.3 implies the Liouville Theorem.

Indeed, since $f(z)$ is a bounded entire function, there exists a positive number M such that $|f(z)| \leq M$ for all $z \in \mathbb{C}$. So the holomorphic function $(1/M)f(z)$ maps $\mathbb{C}$ into $D(0,1)$. We can define a metric λ on $D(0,1)$ with curvature -1. Let $B = 1$ in Theorem 5.3. It follows that $(1/M)f(z)$ is a constant function. Therefore $f(z)$ is a constant function and this proves the Liouville Theorem.

From this, we can see that Theorem 5.3 is a generalization of Liouville Theorem in the form of differential geometry.

According to Liouville Theorem, if an entire function $w = f(z)$ maps $\mathbb{C}$ to a bounded region, then $f(z)$ must be a constant function. If an entire function $w = f(z)$ maps $\mathbb{C}$ to an unbounded region U and the area of $\mathbb{C} \setminus U$ is greater than zero, we can still show that $f(z)$ is a constant. Let c_1 be an interior point of $\mathbb{C}\setminus U$. Let $w_1 = w - c_1$. Then $w_1 = 0$ lies in the complement of $f(\mathbb{C}) - c_1$. Let $w_2 = 1/(w - c_1)$. Then w_2 maps $\mathbb{C}$ to a bounded region. Thus, $w_2 = c_2$. Since $c_2 = 1/(w - c_1)$, w is also a constant.

If an entire function $w = f(z)$ maps $\mathbb{C}$ to an unbounded region U and the area of $\mathbb{C} \setminus U$ is zero ($\mathbb{C} \setminus U$ can be some curves), is $f(z)$ still a constant?

Let us look at the following examples.

Suppose that $w = u + iv = f(z)$ is an entire function and it maps $\mathbb{C}$ to $\mathbb{C} \setminus \{u + i0 | \, 0 \leq u \leq 1\}$. Let

$$w_1 = u_1 + iv_1 = \varphi(w) = \frac{w}{w-1},$$

which maps $\mathbb{C}$ to $\mathbb{C} \setminus \{u_1 + i0 | \, u_1 \leq 0\}$. Let $w_2 = r(w_1) = \sqrt{w_1}$ and choose the principle branch of the square root. Then w_2 maps $\mathbb{C}$ to the right half plane. The Cayley transformation $w_3 = (w_2 - 1)/(w_2 + 1) = s(w_2)$ maps the right half plan to the unit disc. By Liouville Theorem, w_3 is a constant. It follows that w_1, w_2 and w are all constants.

Now we can see that, if an entire function $w = f(z)$ maps $\mathbb{C}$ to an unbounded region, then this function is still a constant even if $\mathbb{C} \setminus U$ is only a line segment. Furthermore, $f(z)$ is a constant even if the length of this segment is arbitrarily small.

How small must the area of $\mathbb{C} \setminus U$ be so that $f(z)$ is not a constant?

Consider another extreme example. The entire function $f(z) = e^z$ maps $\mathbb{C}$ to $U = \mathbb{C} \setminus \{0\}$. So if $\mathbb{C} \setminus U$ is a point, there is a function $f(z)$ that is non-constant. Is an entire function constant if $\mathbb{C} \setminus U$ contains two points? The answer to this question is the Little Picard Theorem.

5.4 The Little Picard Theorem

Theorem 5.4 ***(Little Picard Theorem)*** *Let $w = f(z)$ be an entire function which maps $\mathbb{C}$ to U and $\mathbb{C} \setminus U$ contains at least two points. Then $f(z)$ must be a constant. In other words, a non-constant entire function can have all values of $\mathbb{C}$ except a possible point.*

In order to prove the Little Picard Theorem, we need the following theorem.

Theorem 5.5 *Let U be an open set in $\mathbb{C}$ and $\mathbb{C} \setminus U$ contains at least two points. Then a metric μ can be defined on U such that the curvature $K(z, \mu)$ satisfies*

$$K(z, \mu) \leq -B \leq 0$$

for every point in U, where B is a positive number.

If $\mathbb{C} \setminus U$ contains at least two points, then by Theorem 5.5, a metric μ can be defined on U such that the curvature satisfies $K(z, \mu) \leq -B < 0$

on U, where B is a positive number. By the generalized Liouville Theorem (Theorem 5.3), $f(z)$ must be a constant. Thus, Theorem 5.5 implies Theorem 5.4.

Proof of Theorem 5.5 Choose two points in $\mathbb{C} \setminus U$ and map them to 0 and 1 by a linear transformation. Let $\mathbb{C}_{0,1} = \mathbb{C} \setminus \{0, 1\}$. Define the metric

$$\mu(z) = \frac{(1 + |z|^{1/3})^{1/2}}{|z|^{5/6}} \cdot \frac{(1 + |z-1|^{1/3})^{1/2}}{|z-1|^{5/6}} \tag{5.12}$$

on $\mathbb{C}_{0,1}$. Then $\mu(z)$ is a positive smooth function on $\mathbb{C}_{0,1}$. Now we will calculate the curvature of μ and show that it has negative value.

Since

$$\Delta(\log |z|^{5/6}) = \frac{5}{12}\Delta(\log |z|^2) = 0,$$

we have

$$\begin{aligned}\Delta \log \frac{(1 + |z|^{1/3})^{1/2}}{|z|^{5/6}} &= \frac{1}{2}\Delta \log(1 + |z|^{1/3}) \\ &= 2\frac{\partial}{\partial z}\frac{\partial}{\partial \bar{z}} \log(1 + (z \cdot \bar{z})^{1/6}) = \frac{1}{18|z|^{5/3}(1 + |z|^{1/3})^2}.\end{aligned}$$

Similarly, we get that

$$\Delta \log \left(\frac{(1 + |z-1|^{1/3})^{1/2}}{|z-1|^{5/6}}\right) = \frac{1}{18|z-1|^{5/3}(1 + (z-1)^{1/3})^2}.$$

Thus

$$\begin{aligned}K(z, \mu) = -\frac{1}{18}\Bigg(&\frac{|z-1|^{5/3}}{(1 + |z|^{1/3})^3(1 + |z-1|^{1/3})} \\ &+ \frac{|z|^{5/3}}{(1 + |z|^{1/3})(1 + |z-1|^{1/3})^3}\Bigg).\end{aligned}$$

It follows that

(1) $K(z, \mu) < 0$ for all $z \in \mathbb{C}_{0,1}$;

(2) $\lim_{z \to 0} K(z, \mu) = -\frac{1}{36}$;

(3) $\lim_{z \to 1} K(z, \mu) = -\frac{1}{36}$;

(4) $\lim_{z \to \infty} K(z, \mu) = -\infty$.

So $K(z, \mu)$ has a negative number $-B$ as its upper bound. This proves Theorem 5.5.

The result of Great Picard Theorem is more advanced than Little Picard Theorem. In order to prove the Great Picard Theorem, we need the generalized concept of normal family.

5.5 The Generalization of Normal Family

The concept of normal family was introduced in Section 4.2 and used in the proof of Riemann Mapping Theorem.

Definition 5.1 Let $\{g_i\}$ be a sequence of complex valued functions (not necessarily holomorphic) on a region Ω. If for any given $\varepsilon > 0$ and any compact subset K in Ω, there exists a positive integer J depending only on ε and K, such that

$$|g_j(z) - g(z)| < \varepsilon$$

holds for all $z \in K$ and $j > J$, then we say that $\{g_j\}$ is *normally convergent* on Ω. In other words, if $\{g_j\}$ is uniformly convergent on every compact subset of Ω, then we say that $\{g_j\}$ is normally convergent on Ω.

If for any compact subset K in Ω and any compact subset L in $\mathbb{C}$, there exists a positive integer J which depends only on K and L, such that $g_j(z) \notin L$ for any $z \in K$ and $j > J$, then we say that $\{g_j\}$ is *compactly divergent* on Ω.

In other words, if $\{g_j\}$ diverges uniformly to ∞ on any compact subset of Ω, then we say that $\{g_j\}$ is compactly divergent on Ω.

Definition 5.2 Let $\mathcal{F}$ be a family of complex valued functions on a region Ω. If for any sequence of functions in $\mathcal{F}$, it has either a normally convergent subsequence or a compactly divergent subsequence, then $\mathcal{F}$ is called a *normal family*.

This is a generalization of the definition of normal family in Section 4.2.

Example 5.1 Let $\mathcal{F} = \{f_j\}$, $f_j = z^j$, $j = 1, 2, \cdots$. Then $\mathcal{F}$ is a normal family on $D(0,1)$ since every subsequence normally converges to zero. The family $\mathcal{F}$ is also normal on $\{z|\ |z| > 1\}$ since every subsequence is compactly divergent. The family $\mathcal{F}$ is not normal on the region which contains the points of $|z| = 1$ as its interior points since every subsequence converges to 0 inside the circle and diverges outside the circle.

From the above definition, we have

Theorem 5.6 ***(Montel Theorem)*** *Let $\mathcal{F}$ be a family of holomorphic functions on a region Ω. If for a compact subset K in Ω, there exists a constant M_K such that*

$$|f(z)| \leq M_K \tag{5.13}$$

for every $z \in K$ and every $f \in \mathcal{F}$, then $\mathcal{F}$ is a normal family.

If $|f(z)| \leq M$ for every $z \in \Omega$ and every $f \in \mathcal{F}$ where M is a constant, then the theorem still holds.

Since $\mathcal{F}$ is a family of holomorphic functions and it satisfies (5.13), there are no subsequences which are compactly divergent. Thus, by Theorem 4.3 in Section 4.2 (Montel Theorem), the above theorem holds.

In order to generalize the concept of normal family to the family of meromorphic functions, we use the spherical distance on S^2 instead of the Euclidean distance on $\mathbb{C}$.

Definition 5.3 Let $\mathcal{F}$ be a family of meromorphic functions on a region $\Omega \subset \mathbb{C}^*$. If for any sequence of $\mathcal{F}$ there exists a subsequence which is normally convergent with respect to the sphere distance on Ω, then $\mathcal{F}$ is called a *normal family.*

This definition is the same as the definition of normal family in Section 4.2 except the Euclidean distance is substituted by the spherical distance.

It is easy to see that Definition 5.2 and Definition 5.3 are compatible.

The following Marty Theorem is similar to the Montel Theorem.

Theorem 5.7 ***(Marty Criterion)*** *If $\mathcal{F}$ is a family of meromorphic functions on a region Ω, them $\mathcal{F}$ is a normal family if and only if*

$$\{f^*\sigma |\ f \in \mathcal{F}\} \tag{5.14}$$

is uniformly bounded on any compact subset of Ω, where σ is the spherical metric. In other words, $\mathcal{F}$ is a normal family if and only if for any compact subset K in Ω, there exists a constant M_K such that

$$\frac{2|f'(z)|}{1 + |f(z)|^2} \leq M_K \tag{5.15}$$

for any $z \in K$ and $f \in \mathcal{F}$.

Proof It is easy to see that the uniform boundedness of (5.14) is equivalent to the (5.15).

If (5.15) holds, then

$$d(f(z_1), f(z_2)) = \inf \int_\gamma ds = \inf \int_{\gamma'} \frac{2|f'(z)|}{1+|f(z)|^2}|dz|$$
$$\leq \int_{\gamma_0} \frac{2|f'(z)|}{1+|f(z)|^2}|dz| \leq M_K|z_1 - z_2|,$$

where γ is a curve that join $f(z_1)$ and $f(z_2)$, $\gamma' = f^{-1}(\gamma)$ is contained entirely in K, and γ_0 is a line segment from z_0 to z_1. Thus $\mathcal{F}$ is equicontinuous with respect to the spherical distance. Since the spherical distance is bounded, $\mathcal{F}$ is uniformly bounded. Therefore, by Ascoli-Arzela Theorem (Theorem 4.4 in Section 4.2), $\mathcal{F}$ is a normal family.

Conversely, if (5.15) does not hold, then there exists a compact subset E in Ω and a sequence $\{f_n\}$ in $\mathcal{F}$ such that $\max_{z\in E} f_n^*\sigma(z)$ is unbounded. Since $\mathcal{F}$ is a normal family, there exists a subsequence $\{f_{n_k}\}$ of $\{f_n\}$ such that $f_{n_k} \to f$ as $n_k \to \infty$. For any point of E, there exists a closed disc $\bar{D} \subset \Omega$ such that either f or $1/f$ is holomorphic in Ω. If f is holomorphic, then it is bounded on $\bar{D}$. Since $\{f_{n_k}\}$ is convergent with respect to the spherical distance, the elements of $\{f_{n_k}\}$ have no poles in $\bar{D}$ when n_k is sufficiently large. By Weierstrass Theorem (Theorem 3.1 in Section 3.1), $f_{n_k}^*\sigma$ converges uniformly to $f^*\sigma$ on a disc smaller than $\bar{D}$. Since $f^*\sigma$ is continuous, $f_{n_k}^*\sigma$ is bounded on a smaller disc. Similarly, if $1/f$ is holomorphic, we can show that $(1/f_{n_k})^*\sigma$ is bounded on a smaller disc by the same method. Since $(1/f_{n_k})^*\sigma = f_{n_k}^*\sigma$, we can still get that $f_{n_k}^*\sigma$ is bounded on a smaller disc. Since E is compact, it can be covered by finitely many such disc. Thus, we have that $f_{n_k}^*\sigma$ is bounded on E and this is a contradiction.

The Marty Criterion implies the Montel Theorem.

Theorem 5.8 ***(Montel Theorem)*** *Suppose $\mathcal{F}$ is a family of meromorphic functions, P, Q, R are three different points. If every function of $\mathcal{F}$ assumes its values in $\mathbb{C}^* \setminus \{P, Q, R\}$, then $\mathcal{F}$ is a normal family.*

Proof Map P, Q, R to $0, 1, \infty$ respectively by a fractional linear transformation. Then we only need to show that if no functions in a family of holomorphic functions take value 0 or 1, then this family is normal. In other words, the family of holomorphic functions which assume their values in $\mathbb{C}_{0,1} = \mathbb{C} \setminus \{0, 1\}$ is normal. So we only need to show that $\mathcal{F}$ is a normal family for any disc $D(z_0, \alpha) = \{z \,|\, |z - z_0| < \alpha\} \subseteq \Omega$, and we can assume that $z_0 = 0$.

Let μ be a metric as in Section 5.4. Multiply μ by a constant a (also denoted by μ) such that the upper bound of its curvature is -1. By the

Generalized Ahlfors-Schwarz Lemma (Theorem 5.2 in Section 5.2), we have

$$f^*\mu(z) \leq \lambda_\alpha^A(z)$$

for any $f \in \mathcal{F}$. That is

$$\mu(f(z))\left|\frac{df}{dz}\right| \leq \frac{2\alpha}{\sqrt{A}(\alpha^2 - |z|^2)} \tag{5.16}$$

for any $z \in D(0,\alpha)$.

Compare the spherical metrics $\sigma(w)$ with $\mu(w)$ in $\mathbb{C}_{0,1}$. Obviously,

$$\frac{\sigma(w)}{\mu(w)} = \frac{\dfrac{2}{1+|w|^2}}{\dfrac{a(1+|w|^{1/3})^{1/2}(1+|w-1|^{1/3})^{1/2}}{|w|^{5/6}|w-1|^{5/6}}} \to 0$$

when w tends to $0, 1$ or ∞. Thus, there exists a positive number M such that $\sigma(w) \leq M\mu(w)$. By (5.16)

$$\begin{aligned} f^*\sigma(z) = \sigma(f(z))\left|\frac{df}{dz}\right| &\leq M\mu(f(z))\left|\frac{df}{dz}\right| \\ = Mf^*\mu(z) \leq M\lambda_\alpha^A &= \frac{2\alpha M}{\sqrt{A}(\alpha^2 - |z|^2)} \end{aligned}$$

when $z \in D(0,\alpha)$. Therefore, $f^*\sigma$ is bounded on the compact subsets of $D(0,\alpha)$ and the bound is independent of $f \in \mathcal{F}$. By Marty Criterion (Theorem 5.7), $\mathcal{F}$ is a normal family.

By the proof of Theorem 5.8, we also get

Theorem 5.9 ***(Montel Theorem)*** *Let $\mathcal{F}$ be a family of holomorphic functions on a region Ω. If every function in $\mathcal{F}$ does not assume the same two values, then $\mathcal{F}$ is a normal family.*

5.6 The Great Picard Theorem

Recall the Weierstrass Theorem (Theorem 3.3 in Section 3.2): If $f(z)$ is holomorphic in $D'(0,r) = D(0,r) \setminus \{0\}$ and $z = 0$ is an essential singularity of $f(z)$, then the values of $f(z)$ are dense in $\mathbb{C}$

The Great Picard Theorem gives a more advanced characterization about the value distribution of a function around its essential singularity.

Theorem 5.10 ***(Great Picard Theorem)*** *If $f(z)$ is holomorphic in $D'(0,1)$ and $z = 0$ is an essential singularity, then in each neighborhood of $z = 0$, f assumes every number in $\mathbb{C}$ with one possible exception. The assertion still holds if $z = 0$ is substituted by any other point.*

The Great Picard Theorem is a generalization of the Little Picard Theorem and it deepens the Weierstrass Theorem.

According to the discussion of Section 3.3, if $f(z)$ is an entire function and the infinity is a pole, then $f(z)$ is a polynomial. By the Fundamental Theorem of Algebra (Theorem 2.12 in Section 2.4), the value of $f(z)$ can be any number in $\mathbb{C}$. If the infinity is a removable singularity of $f(z)$, then $f(z)$ is a bounded entire function. By Liouville Theorem, $f(z)$ must be a constant. If the infinity is an essential singularity of $f(z)$, then by the Great Picard Theorem, $f(z)$ assumes each number in $\mathbb{C}$ near the infinity with only one possible exception. This implies the Little Picard Theorem.

Hence, the Little Picard Theorem is a corollary of the Great Picard Theorem.

Now we prove the Great Picard Theorem using the results from the last section.

Proof of Theorem 5.10 Suppose the contrary. Without loss of generosity, assume that $f(z)$ is holomorphic on $D'(0,1)$, and that the image region of $D'(0,1)$ under f does not contain 0 and 1. We need to show that $z = 0$ must be a removable singularity or a pole to get the contradiction.

Define

$$f_n(z) = f\left(\frac{z}{n}\right), \quad 0 < |z| < 1.$$

Let $\mathcal{F} = \{f_n\}$. Then every f_n assumes its values in $\mathbb{C}_{0,1}$. By Theorem 5.9, $\mathcal{F}$ is a normal family. Thus, there exists a subsequence $\{f_{n_k}\}$ in $\{f_n\}$ which is either normally convergent or compactly divergent. If $\{f_{n_k}\}$ is normally convergent, then $\{f_{n_k}\}$ converges uniformly on every compact subset of $D'(0,1)$. It follows that $\{f_{n_k}\}$ is bounded and especially has a bound M on $\{z \,|\, |z| = 1/2\}$ and f has a bound M on $\{z \,|\, |z| = 1/2n_k\}$. By the Maximal Modulus Principle, f has the bound M on $0 < |z| < 1/2$. Thus $z = 0$ is a removable singularity of $f(z)$ (Theorem 9 in Section 2.3)

If $\{f_{n_k}\}$ is compactly divergent, then by the same method we can prove that $1/f \to 0$ as $z \to 0$. Equivalently, $f \to \infty$ as $z \to 0$. Therefore, $z = 0$ is a pole of $f(z)$ and the theorem follows.

The Great Picard Theorem and the Little Picard Theorem are the important theorems in complex analysis especially in the theory of value distribution. Regular textbooks for undergraduate students do not include these theorems since the elliptic modular function, which is considered too advanced for the undergraduates, is needed in the proofs. There are a few simplified proofs. In this chapter, we chose the proof from the point of view of differential geometry. In 1938, L. V. Ahlfors established the important Ahlfors-Schwarz Lemma (Theorem 5.1). In 1939, R. M. Robinson followed this idea and proved Picard Theorem by the method of differential geometric instead of using the elliptic modular function. Later on, there are some improvements such as the work of H. Grauert and H. Reckziegel [1], Z. Kobayashi [1], L. Zalcman [1], D. Minda and G.Schober [1] and S. G. Krantz. These are the references of this chapter especially the work of Minda, Schober and Krantz. Our way of writing this part not only simplified the proof of the Picard Theorem but also gives the reader an introduction of how to treat the problems of complex analysis by the method of differential geometry. Moreover, the method of differential geometry in the proofs of the Great Picard Theorem and the Little Picard Theorem can also be used in the proofs of some other important theorems of complex analysis such as the Bloch Theorem, the Landau Theorem, and the Schottky Theorem, etc. We now give the statements of these three theorems without proofs. The reader can refer to the works mentioned above and L. V. Ahlfors [2], [3], J. B. Conway [1].

Theorem 5.11 ***(Bloch Theorem)*** *If $f(z)$ is holomorphic on the unit disc D and $f'(0) = 1$, then $f(D)$ contains a disc with radius B where B is a positive number independent of f.*

Theorem 5.12 ***(Landau Theorem)*** *If $f(z) = a_0 + a_1 z + \cdots$ $(a_1 \neq 0)$ is a holomorphic function on $D(0, r)$ and omits the values 0 and 1, then $r \leq R(a_0, a_1)$ where $R(a_0, a_1)$ is a constant depending only on a_0 and a_1.*

Theorem 5.13 ***(Schottky Theorem)*** *If $f(z) = a_0 + a_1 z + \cdots$ is a holomorphic function in $D(0, r)$ and omits the values 0 and 1, then for every $\theta \in (0, 1)$, there exists a constant $M(a_0, \theta)$ depending only on a_0 and θ such that $|f(z)| \leq M(a_0, \theta)$ for all $|z| \leq \theta r$.*

It can also be proved that the Bloch Theorem implies the Little Picard Theorem and the Schottky Theorem implies the Great Picard Theorem.

In the process of proving the Little Picard Theorem and the Great Picard Theorem, a critical step is the construction of the metric μ. The curvature of this metric has a negative upper bound. There exists a positive number M such that $\sigma \leq M\mu$. The metric defined by (5.12) is not the only one which satisfies the above conditions. It is possible to construct other metrics which also satisfy the above requirement. Interested reader can refer to R. M. Robinson [1].

Exercise V

1. Show that

$$\Delta = \frac{\partial^2}{\partial r^2} + \frac{1}{r}\frac{\partial}{\partial r} + \frac{1}{r^2}\frac{\partial^2}{\partial \theta^2},$$

where $z = re^{i\theta}$.

2. Prove Theorem 5.2.

3. Prove that the Euclidean metric is an invariant under the group of Euclidean motions.

4. Let $P = (x_1, x_2, x_3)$ and $P' = (x_1', x_2', x_3')$ be two points on the Riemann sphere. Show that the length of the arc PP' on the great circle passing through P and P' is

$$d(P, P') = 2\arctan\sqrt{\frac{1 - x_1x_1' - x_2x_2' - x_3x_3'}{1 + x_1x_1' + x_2x_2' + x_3x_3'}}.$$

If z, z' are the corresponding points of P, P' on $\mathbb{C}$ under the stereographic projection, then by (5.5)

$$d(z, z') = 2\arctan\left|\frac{z - z'}{1 + z\bar{z}'}\right|.$$

Moreover, show that the metric is

$$ds^2 = \frac{4|dz|^2}{(1 + |z|^2)^2}.$$

5. Verify that the curvature of the Euclidean metric is 0, the curvature of Poincaré metric is -1 and the curvature of the spherical metric is $+1$.

6. Verify the correctness of the Little Picard Theorem for the function $f(z) = e^z$.

7. What is the exceptional value of $e^z + 1$?

8. Show that the functions $\cosh z$ and $\sinh z$ assume all complex values.

9. Verify the Great Picard Theorem for the function $e^{1/z}$ near $z = 0$.

Appendix III Curvature

Detailed treatments about the curvature of surface in three dimensional Euclidean space can be found in any undergraduate textbook of Differential Geometry. We give a simple introduction here for convenience. The reader can skip this appendix if he/she has learned this concept before.

Let $D \subset \mathbb{R}^2$ be a region and $(u,v) \in D$. If

$$\vec{r}(u,v) = (x(u,v), y(u,v), z(u,v)) \in \mathbb{R}^3,$$

where x, y, z have second order continuous partial derivatives with respect to u, v and $\vec{r}_u \times \vec{r}_v \neq 0$ (where $\vec{r}_u$ and $\vec{r}_v$ are the partial derivatives of $\vec{r}$ with respect to u and v), then $S = \vec{r}(D)$ is a surface in $\mathbb{R}^3$. For $(u,v) \in D$, $\vec{r}(u,v)$ is a point P on S and $\vec{r}_u$ and $\vec{r}_v$ form a basis of the tangent plane, T_PS, of S at the point P. The *First Fundamental Form* of surface S is

$$I = E(du)^2 + 2Fdudv + G(dv)^2 = ds^2,$$

where $E = \vec{r}_u \cdot \vec{r}_u$, $F = \vec{r}_u \cdot \vec{r}_v$ and $G = \vec{r}_v \cdot \vec{r}_v$. This represents the square of the length of the infinitesimal vector $d\vec{r} = \vec{r}_u du + \vec{r}_v dv$.

Let $Q = \vec{r}(u + \Delta u, v + \Delta v)$ be a point near P. Then

$$\begin{aligned}\overrightarrow{PQ} &= \vec{r}(u + \Delta u, v + \Delta v) - \vec{r}(u,v) \\ &= \vec{r_u}\Delta u + \vec{r_v}\Delta v + \frac{1}{2}(\vec{r_{uu}}(\Delta u)^2 + 2\vec{r_{uv}}\Delta u \Delta v + \vec{r_{vv}}(\Delta v)^2) + \cdots,\end{aligned}$$

where $\vec{r}_{uu}$, $\vec{r}_{uv}$ and $\vec{r}_{vv}$ are the second order partial derivative of $\vec{r}$ with respect to u, the second order partial derivative with respect to u and v and the second order partial derivative with respect to v respectively. Thus, the directed distance from point Q to T_PS is

$$\delta = \overrightarrow{PQ} \cdot \vec{n} = \frac{1}{2}(\vec{r}_{uu}(\Delta u)^2 + 2\vec{r}_{uv}\Delta u \Delta v + \vec{r}_{vv}(\Delta v)^2) + \cdots,$$

where

$$\vec{n} = \frac{\vec{r}_u \times \vec{r}_v}{|\vec{r}_u \times \vec{r}_v|}$$

is a normal vector of S at the point P. The omitted parts in the above equation are higher order terms of Δu and Δv. Choose the principle part of 2δ (the expression of 2δ with the higher degree terms omitted) as the *Second Fundamental Form* of the surface S

$$II = L(du)^2 + 2Mdudv + N(dv)^2,$$

where

$$L = \frac{\vec{r}_{uu} \cdot \vec{r}_u \times \vec{r}_v}{|\vec{r}_u \times \vec{r}_v|}, \quad M = \frac{\vec{r}_{uv} \cdot \vec{r}_u \times \vec{r}_v}{|\vec{r}_u \times \vec{r}_v|}, \quad N = \frac{\vec{r}_{vv} \cdot \vec{r}_u \times \vec{r}_v}{|\vec{r}_u \times \vec{r}_v|}.$$

Since $\vec{r}_u \cdot \vec{n} = 0$ and $\vec{r}_v \cdot \vec{n} = 0$, taking derivatives with respect to u and v, we can get that

$$\vec{r}_{uu} \cdot \vec{n} + \vec{r}_u \cdot \vec{n}_u = 0, \quad \vec{r}_{uv} \cdot \vec{n} + \vec{r}_u \cdot \vec{n}_v = 0,$$

$$\vec{r}_{uv} \cdot \vec{n} + \vec{r}_v \cdot \vec{n}_u = 0, \quad \vec{r}_{vv} \cdot \vec{n} + \vec{r}_v \cdot \vec{n}_v = 0.$$

It follows that $II = -d\vec{r} \cdot d\vec{n}$. The second fundamental form indicates how the surface curved near the point P.

The important facts are, both form I and form II are geometric invariants. That is, if another set of parameters $(\bar{u}, \bar{v})$ is used to represent the surface S, then these two forms are still the same. The fundamental theorem in the theory of curves is: If any two differentiable functions $f(x)(> 0)$ and $g(x)$ are given, then there exists one and only one curve with curvature f and torsion g when the initial conditions are given. In the theory of surfaces, if any six functions which satisfy certain conditions (Gauss equation and Codazzi equation) are given, then there must exists a surface such that the first and the second fundamental forms are defined by these six functions and the surface is unique up to a rigid motion. This is the fundamental theorem of the surface theory.

Suppose C is a curve $\vec{r} = \vec{r}(s)$ on the surface S that passes through the point P, where s is the arc length, then

$$\vec{T}(s) = \frac{d\vec{R}}{ds} = \vec{r}_u \frac{du}{ds} + \vec{r}_u \frac{du}{ds} + \vec{r}_v \frac{dv}{ds}$$

is a unit tangent vector of S at point P, and

$$T'(s) = \frac{d^2\vec{r}}{ds^2} = \vec{r}_{uu} \left(\frac{du}{ds}\right)^2 + 2\vec{r}_{uv} \left(\frac{du}{ds}\right) \left(\frac{dv}{ds}\right) + \vec{r}_{vv} \left(\frac{dv}{ds}\right)^2 + \vec{R}$$

is the normal vector which is perpendicular to $\vec{T}(s)$, where $\vec{R}$ is a tangent vector of S at point P. If $\vec{N}(s)$ is a unit normal vector of the curve C at the point P, then $T'(s) = k\vec{N}(s)$. Thus, we call

$$k_n = k\vec{N} \cdot \vec{n} = \vec{r}_{uu} \left(\frac{du}{ds}\right)^2 + 2\vec{r}_{uv} \left(\frac{du}{ds}\right) \left(\frac{dv}{ds}\right) + \vec{r}_{vv} \left(\frac{dv}{ds}\right)^2$$

the *normal curvature* of the curve C at the point P, where $\vec{n}$ is the unit normal vector of surface S at point P. Obviously,

$$k_n = \frac{Ldu^2 + 2Mdudv + Ndv^2}{Edu^2 + 2Fdudv + Gdv^2},$$

is the quotient of the second fundamental form and the first fundamental form. Since k_n depends on the direction of du, dv, the directions in which k_n reaches its extreme values (maximal value or minimal value) are called the *principle directions* of the surface at the point. The extreme values are called the *principle curvatures* of the surface at the point. In general, two kind of curvatures can be defined at any point on a surface. One is called *Gaussian curvature* (or *total curvature*) K, the product of the two principle curvatures. The other is called the *mean curvature* H, the arithmetic mean of two principle curvatures. Now we find the expressions of these two curvatures in terms of E, F, G, L, M and N.

Obviously,

$$\begin{pmatrix} E & F \\ F & G \end{pmatrix}$$

is a positive definite matrix, and it is easy to prove that

(1) The equation

$$\det\left(\lambda\begin{pmatrix} E & F \\ F & G \end{pmatrix} - \begin{pmatrix} L & M \\ M & N \end{pmatrix}\right) = 0$$

has two real roots λ_1 and λ_2;

(2) If $\lambda_1 \neq \lambda_2$, then there exists

$$A = \begin{pmatrix} a_{11} & a_{12} \\ a_{21} & a_{22} \end{pmatrix}$$

such that

$$A\begin{pmatrix} E & F \\ F & G \end{pmatrix}A' = \begin{pmatrix} 1 & 0 \\ 0 & 1 \end{pmatrix}$$

and

$$A\begin{pmatrix} L & M \\ M & N \end{pmatrix}A' = \begin{pmatrix} \lambda_1 & 0 \\ 0 & \lambda_2 \end{pmatrix}$$

where A' is the transpose of A.

Let $(d\bar{u}, d\bar{v}) = (du, dv)A^{-1}$. Then

$$k_n = \frac{\lambda_1(d\bar{u})^2 + \lambda_2(d\bar{v})^2}{(d\bar{u})^2 + (d\bar{v})^2}.$$

Let $\bar{\lambda} = \max(\lambda_1, \lambda_2)$ and $\underline{\lambda} = \min(\lambda_1, \lambda_2)$. Then

$$\begin{aligned} k_n &= \bar{\lambda} - (\bar{\lambda} - \lambda_1)\frac{(d\bar{u})^2}{(d\bar{u})^2 + (d\bar{v})^2} - (\bar{\lambda} - \lambda_2)\frac{(d\bar{v})^2}{(d\bar{u})^2 + (d\bar{v})^2} \\ &= \underline{\lambda} + (\lambda_1 - \underline{\lambda})\frac{(d\bar{u})^2}{(d\bar{u})^2 + (d\bar{v})^2} + (\lambda_2 - \underline{\lambda})\frac{(d\bar{v})^2}{(d\bar{u})^2 + (d\bar{v})^2}. \end{aligned}$$

Hence, $\underline{\lambda} \le k_n \le \bar{\lambda}$ for any direction du, dv. It can be shown that there exist some directions that have λ_1, λ_2 as their normal curvatures. So λ_1, λ_2 are the principle curvatures. Therefore, $K = \lambda_1\lambda_2$ and $H = (\lambda_1 + \lambda_2)/2$. By (1), we have

$$K = \frac{LN - M^2}{EG - F^2}, \quad H = \frac{1}{2} \cdot \frac{GL - 2FM + EN}{EG - F^2}.$$

The above equations also holds when $\lambda_1 = \lambda_2$.

The most important subject discussed in differential geometry is the curvature, especially the Gaussian curvature. In general, when a curvature is mentioned, we mean the Gaussian curvature.

Next, we find the Gaussian curvature when the first fundamental form or the metric is

$$ds^2 = \rho^2 du^2 + \rho^2 dv^2.$$

Since

$$E = G = \vec{r}_u \cdot \vec{r}_u = \rho^2, \quad F = \vec{r}_u \cdot \vec{r}_v = 0,$$

we have

$$|\vec{r}_u \times \vec{r}_v|^2 = (\vec{r}_u \cdot \vec{r}_u)(\vec{r}_v \cdot \vec{r}_v) - (\vec{r}_u \cdot \vec{r}_v) = \rho^2,$$

and $|\vec{r}_u \times \vec{r}_v| = \rho$. We also have

$$LN = \frac{1}{\rho^4}\begin{vmatrix} \vec{r}_{uu} \cdot \vec{r}_{vv} & \vec{r}_{uu} \cdot \vec{r}_u & \vec{r}_{uu} \cdot \vec{r}_v \\ \vec{r}_u \cdot \vec{r}_{vv} & \rho^2 & 0 \\ \vec{r}_v \cdot \vec{r}_{vv} & 0 & \rho^2 \end{vmatrix},$$

$$M^2 = \frac{1}{\rho^4} \begin{vmatrix} \vec{r}_{uv} \cdot \vec{r}_{uv} & \vec{r}_{uv} \cdot \vec{r}_u & \vec{r}_{uv} \cdot \vec{r}_v \\ \vec{r}_u \cdot \vec{r}_{uv} & \rho^2 & 0 \\ \vec{r}_v \cdot \vec{r}_{uv} & 0 & \rho^2 \end{vmatrix}.$$

Take the partial derivatives of the equations

$$\vec{r}_u \cdot \vec{r}_u = \vec{r}_v \cdot \vec{r}_v = \rho^2, \quad \vec{r}_u \cdot \vec{r}_v = 0$$

with respect to u and v, we get

$$\vec{r}_{uu} \cdot \vec{r}_u = \rho\rho_u, \quad \vec{r}_{uv} \cdot \vec{r}_u = \rho\rho_v, \quad \vec{r}_{uv} \cdot \vec{r}_v = \rho\rho_v,$$

$$\vec{r}_{vv} \cdot \vec{r}_v = \rho\rho_v, \quad \vec{r}_{uu} \cdot \vec{r}_v = -\rho\rho_v, \quad \vec{r}_u \cdot \vec{r}_{uv} = -\rho\rho_u$$

and

$$\frac{\partial}{\partial v}(\vec{r}_{uu} \cdot \vec{r}_v) = \vec{r}_{uuv} \cdot \vec{r}_v + \vec{r}_{uu} \cdot \vec{r}_{vv} = -\rho_v^2 - \rho\rho_{vv},$$

$$\frac{\partial}{\partial u}(\vec{r}_{uu} \cdot \vec{r}_v) = \vec{r}_{uuv} \cdot \vec{r}_v + \vec{r}_{uv} \cdot \vec{r}_{uv} = \rho_u^2 + \rho\rho_{uu}.$$

It follows that

$$\vec{r}_{uu} \cdot \vec{r}_{vv} - \vec{r}_{uv} \cdot \vec{r}_{uv} = -(\rho_u^2 + \rho_v^2) - \rho\Delta\rho.$$

Substitute these results into the determinant expressions of LN and M^2, we get that

$$\begin{aligned} LN - M^2 &= \frac{1}{\rho^4} \begin{vmatrix} \vec{r}_{uu} \cdot \vec{r}_{vv} & \rho\rho_u & -\rho\rho_v \\ -\rho\rho_u & \rho^2 & 0 \\ \rho\rho_v & 0 & \rho^2 \end{vmatrix} - \frac{1}{\rho^4} \begin{vmatrix} \vec{r}_{uv} \cdot \vec{r}_{uv} & \rho\rho_v & \rho\rho_u \\ \rho\rho_v & \rho^2 & 0 \\ \rho\rho_u & 0 & \rho^2 \end{vmatrix} \\ &= \vec{r}_{uu} \cdot \vec{r}_{vv} + \rho_u^2 + \rho_v^2 - \vec{r}_{uv} \cdot \vec{r}_{uv} + \rho_u^2 + \rho_v^2 \\ &= \rho_u^2 + \rho_v^2 - \rho\Delta\rho, \end{aligned}$$

Since $EG - F^2 = \rho^4$, the Gaussian curvature is

$$\frac{\rho_u^2 + \rho_v^2}{\rho^4} - \frac{\Delta\rho}{\rho^3} = -\frac{1}{\rho^2}\Delta \log \rho.$$

Chapter 6

A First Taste of Function Theory of Several Complex Variables

6.1 Introduction

For a better understanding of complex analysis, in this chapter, we will discuss some essential differences between the theory of complex functions in one variable and the theory of complex functions in several variables.

Just as the generalizations of other mathematical theories to higher dimensions, parts of complex analysis can be generalized without any difficulties. However other parts have to be developed in a separate way. We will pay close attention to the later.

Similar to complex analysis, there is a long history of the studying of the theory of complex functions in several variables. In the beginning of the last century, the discovery of Poincaré Theorem and Hartogs Theorem started a new era of the function theory of several complex variables.

Let $\mathbb{C}^n$ be the n-dimensional complex Euclidean space and $z = (z_1, z_2, \cdots, z_n) \in \mathbb{C}^n$. A *unit ball* is defined as $B(0,1) = \{z \in \mathbb{C}^n | \, |z_1|^2 + \cdots + |z_n|^2 < 1\}$. A *polydisc* is defined as $D^n(0.1) = \{z \in \mathbb{C}^n | \, |z_1| < 1, \cdots, |z_n| < 1\}$. A map $f(z) = (f_1(z), f_2(z), \cdot, f_n(z))$ is holomorphic on a region $\Omega \subseteq \mathbb{C}^n$ if every $f_j(z)$ $(j = 1, 2, \cdots, n)$ is holomorphic on Ω. A function $g(z) = g(z_1, z_2, \cdots, z_n)$ is holomorphic on Ω if $g(z)$ is holomorphic with respect to any one of z_i while the other $n-1$ variables are fixed. A function f is a *biholomorphic* mapping if it is one-to-one, onto, and f^{-1} is also holomorphic (the last requirement can be omitted but the proof of this fact is complicated).

The Poincaré Theorem tells us that: the Riemann Mapping Theorem (in Chapter 4) does not hold in higher dimensions (≥ 2). In other words, there are no biholomorphic functions which map a unit ball in $\mathbb{C}^n$ $(n \geq 2)$ to a polydisc.

Hartogs Theorem tells us that, there exists a region Ω in $\mathbb{C}^n$ ($n \geq 2$) such that if a function f is holomorphic on Ω, then f is also holomorphic on a region larger than Ω. In other words, there exists a region Ω in $\mathbb{C}^n$ ($n \geq 2$) such that if a function f is holomorphic on Ω, then f has an analytic continuation on a region larger than Ω. This assertion does not exists in complex analysis of one variable. So Hartogs Theorem brings the attention to the importance of regions in the studying of the function theory.

The effort of mathematicians in more than two hundred years gives rise to the maturity of complex analysis in one variable. There is still a long way to go for the theory of complex functions in several variables.

As a starting point of the function theory of several complex variables, in this chapter, we will study some generalized results from complex analysis and prove the Poincaré Theorem and the Hartogs Theorem. Only the case of two complex variables is considered here for simplicity. In other words, our discussion is on $\mathbb{C}^2 = \mathbb{C} \times \mathbb{C}$. A special case of Hartogs Theorem will be proved.

The proofs of the following theorems is omitted since they are similar to the one dimensional case.

Theorem 6.1 ***(Cauchy Integral Formula)*** *Let $w = (w_1, w_2) \in \mathbb{C}^2$ and $r > 0$. Let*

$$D^2(w,r) = \{z = (z_1, z_2) \in \mathbb{C}^2 |\ |z_1 - w_1| < r, |z_2 - w_2| < r\}$$

be a polydisc with radius r and center w. If $f(z)$ is holomorphic on $\bar{D}(w,r)$, then

$$f(z) = \frac{1}{(2\pi i)^2} \int_{\partial D(w_1,r)} \int_{\partial D(w_2,r)} \frac{f(\zeta_1, \zeta_2)}{(\zeta_1 - z_1)(\zeta_2 - z_2)} d\zeta_1 d\zeta_2 \tag{6.1}$$

for any $z \in D^2(w,r)$.

This can be proved by applying Cauchy Integral Formula of one variable twice.

Similar to the case of one variable, the n-th partial derivative of $f(z)$ can be found by (6.1) for any positive integer n. Moreover,

$$\left(\frac{\partial}{\partial z_i}\right)^j \left(\frac{\partial}{\partial z_2}\right)^k f(z) = \frac{j!k!}{(2\pi i)^2} \int_{\partial D(w_1,r)} \int_{\partial D(w_2,r)} \frac{f(\zeta_1, \zeta_2)}{(\zeta_1 - z_1)^{j+1}(\zeta_2 - z_2)^{k+1}} d\zeta_1 d\zeta_2,$$

where j, k are any non-negative integers. This implies the Cauchy inequality

$$\left|\left(\frac{\partial}{\partial z_1}\right)^j\left(\frac{\partial}{\partial z_2}\right)^k f(z)\right| \le \frac{Mj!k!}{r^{j+k}},$$

where

$$M = \sup_{|\zeta_1 - z_1|=r,\,|\zeta_2 - z_2|=r} |f(\zeta)|$$

and $\xi = (\xi_1, \xi_2)$.

Similar to the case of one variable, if $f(z)$ is holomorphic on a neighborhood of $\bar{D}^2(w,r)$, then $f(z)$ has the Taylor expansion

$$f(z) = \sum_{j,k=0}^{\infty} a_{jk}(z_1 - w_1)^j (z_2 - w_2)^k,$$

and it converges absolutely and uniformly on $\bar{D}^2(w,r)$, where

$$a_{jk} = \frac{1}{j!k!}\left(\frac{\partial}{\partial z_1}\right)^j\left(\frac{\partial}{\partial z_2}\right)^k f(w).$$

Also, similar to the case of one variable, we can prove that

Theorem 6.2 *If $f(z)$ is holomorphic on a region $U \subseteq \mathbb{C}^2$ and vanishes on an open subset of U, then $f(z)$ is identically equal to zero on U.*

Theorem 6.3 ***(Maximal Modulus Principle)*** *Let U be a bounded region in $\mathbb{C}^2$, $f(z)$ be a holomorphic function on U and*

$$M = \sup_{\zeta \in \partial U} \lim_{\substack{z \to \zeta \\ z \in U}} |f(z)|.$$

Then $|f(z)| < M$ for all $z \in U$ except when f is a constant.

Similar to the Weierstrass Theorem in Section 3.1 we have

Theorem 6.4 ***(Weierstrass Theorem)*** *Let Ω be a region in $\mathbb{C}^2$, $\{f_n\}$ be a sequence of holomorphic functions on Ω that converges uniformly on every compact subset of Ω. Then $f = \lim_{n\to\infty} f_n$ is holomorphic on Ω and the sequence*

$$\left\{\left(\frac{\partial}{\partial z_1}\right)^j\left(\frac{\partial}{\partial z_2}\right)^k f_n\right\}$$

converges uniformly to

$$\left(\frac{\partial}{\partial z_1}\right)^j \left(\frac{\partial}{\partial z_2}\right)^k f$$

on every compact subset of Ω.

Similar to the Montel Theorem in Section 4.2 we have

Theorem 6.5 *(Montel Theorem)* *If $\mathcal{F} = \{f_n\}$ is a family of holomorphic functions on $\Omega \subseteq \mathbb{C}^2$ and there exists an $M > 0$ such that $|f(z)| \leq M$ for all $z \in \Omega$ and $f \in \mathcal{F}$, then $\mathcal{F}$ is a normal family.*

6.2 Cartan Theorem

We determined the group of holomorphic automorphisms of the unit disc in Section 2.5. In this section, we will determine the group of holomorphic automorphisms of the unit ball and the polydisc of $\mathbb{C}^2$ and give a proof of Poincaré Theorem.

We determine the group of holomorphic automorphisms of the unit disc based on Schwarz Lemma. In the case of several complex variables, we need to use the generalized Schwarz Lemma which is equivalent to the following two theorems by Cartan.

Theorem 6.6 *(Cartan Theorem)* *Let $U \subseteq \mathbb{C}^2$ be a bounded region and $P \in U$. If $f = (f_1, f_2)$ is a holomorphic function which maps U into U with $f(P) = P$ and $J_f(P) = I$, then $f(z) \equiv z$. Here $J_f(z)$ is the Jacobi matrix of f at point z*

$$\begin{pmatrix} \dfrac{\partial f_1}{\partial z_1} & \dfrac{\partial f_1}{\partial z_2} \\ \dfrac{\partial f_2}{\partial z_1} & \dfrac{\partial f_2}{\partial z_2} \end{pmatrix}$$

and I is the unit matrix

$$\begin{pmatrix} 1 & 0 \\ 0 & 1 \end{pmatrix}.$$

Proof Without loss of generosity, let $P = 0$. Suppose the contrary. Then $f(z)$ has the Taylor expansion at point 0

$$f(z) = z + A_m(z) + \cdots ,$$

where $A_m(z) = (A_m^{(1)}(z), A_m^{(2)}(z))$ is the first non-zero term and $A_m^{(1)}(z)$, $A_m^{(2)}(z)$ are homogeneous polynomials with degree $m\,(\geq 2)$.

Let $f^1 = f$, $f^2 = f \circ f$, $\cdots$, $f^j = f^{j-1} \circ f$, $(j \geq 2)$. Then

$$f^1(z) = z + A_m(z) + \cdots ,$$

$$\begin{aligned} f^2(z) &= f(z) + A_m(f(z)) + \cdots \\ &= z + A_m(z) + A_m(z) + \cdots \\ &= z + 2A_m(z) + \cdots , \end{aligned} \tag{6.2}$$

$$\cdots\cdots\cdots\cdots\cdots\cdots\cdots\cdots\cdots$$

$$f^j(z) = z + jA_m(z) + \cdots .$$

Since U is a bounded region and f maps U into U, by Theorem 6.5 (Montel Theorem) $\{f^j\}$ is a normal family. In other words, there exists a subsequence $\{f^{j_l}\}$ such that $f^{j_l} \to F$ as $j_l \to \infty$. By Theorem 6.4, the value of the m-th derivative of f^{j_l} at point 0 tends to the m-th derivative of F at 0. By (6.2) the m-th derivative of f^{j_l} at 0 tends to ∞ as $j_l \to \infty$. On the other hand, the value of the m-th derivative of F at point 0 can not be ∞. This is a contradiction. Therefore $A_m(z) \equiv 0$ on U and it follows that $f(z) \equiv z$.

In the case of one complex variable, U becomes a unit disc and this theorem becomes: If a holomorphic function $f(z)$ maps D to D, $f(0) = 0$, $f'(0) = 1$, then $f(z) \equiv z$. This is the equality in the Schwarz Lemma.

Now we prove the other Cartan Theorem.

Let $U \subseteq \mathbb{C}^2$ be a region. If for any point $(z_1, z_2) \in U$ and any complex number μ with $|\mu| < 1$, the point $(\mu z_1, \mu z_2) \in U$, then U is called a *circular domain*.

Theorem 6.7 ***(Cartan Theorem)*** *Let $U \subset \mathbb{C}^2$ be a bounded circular domain, f be a biholomorphic function which maps U to U and $f(0) = 0$. Then f is a linear map. That is, $f(z) = zA$, where A is a matrix with constant elements.*

Proof Let $\theta \in [0, 2\pi]$ and $\rho_0(z_1, z_2) = (e^{i\theta} z_1, e^{i\theta} z_2)$. Assume that $g = \rho_{-\theta} \circ f^{-1} \circ \rho_\theta \circ f$. Then

$$J_g(0) = \begin{pmatrix} e^{-i\theta} & 0 \\ 0 & e^{-i\theta} \end{pmatrix} J_f^{-1}(0) \begin{pmatrix} e^{i\theta} & 0 \\ 0 & e^{i\theta} \end{pmatrix} J_j(0) = I.$$

Since $g(z)$ maps U to U with $g(0) = 0$ and $J_g(0) = I$, by Theorem 6.6, we have $g(z) \equiv z$, or equivalently

$$f \circ \rho_\theta = \rho_\theta \circ f. \tag{6.3}$$

Expand f to a convergent power series near point 0

$$f(z) = \sum_{j,k=0}^{\infty} a_{jk} z_1^j z_2^k = \left(\sum_{j,k=0}^{\infty} a_{jk}^{(2)} z_1^j z_2^k, \sum_{j,k=0}^{\infty} a_{jk}^{(2)} z_1^j z_2^k \right),$$

then

$$\rho_\theta \circ f = (e^{i\theta} f_1, e^{i\theta} f_2) = \left(\sum_{j,k=0}^{\infty} a_{jk}^{(1)} e^{i\theta} z_1^j z_2^k, \sum_{j,k=0}^{\infty} a_{jk}^{(2)} e^{i\theta} z_1^j z_2^k \right),$$

and

$$\begin{aligned} f \circ \rho_\theta &= \left(\sum_{j,k=0}^{\infty} a_{jk}^{(1)} (e^{i\theta} z_1)^j (e^{i\theta} z_2)^k, \sum_{j,k=0}^{\infty} a_{jk}^{(2)} (e^{i\theta} z_1)^j (e^{i\theta} z_2)^k \right) \\ &= \left(\sum_{j,k=0}^{\infty} a_{jk}^{(1)} e^{i(j+k)\theta} z_1^j z_2^k, \sum_{j,k=0}^{\infty} a_{jk}^{(2)} e^{i(j+k)\theta} z_1^j z_2^k \right). \end{aligned}$$

Comparing the coefficients of the above two equations, we have that every $a_{jk} = 0$ except the terms with $j + k = 1$. It follows that $f(z)$ is linear near the point 0. Therefore, $f(z)$ is linear on the entire U by Theorem 6.2.

In the case of one complex variable, U becomes the unit disc D and this theorem becomes: If a univalent holomorphic function $f(z)$ maps D to D and $f(0) = 0$, then $f(z) = e^{i\theta} z$.

Now we can determine the group of holomorphic automorphisms of the unit ball and the polydisc on $\mathbb{C}^2$ by these two theorems.

6.3 Groups of Holomorphic Automorphisms of The Unit Ball and The Bidisc

Let U be a region in $\mathbb{C}^2$. If there exists a biholomorphic function $f(z)$ which maps U to U, then $f(z)$ is called a *holomorphic automorphism* or *biholomorphic automorphism* on U. The group formed by all holomorphic automorphisms on U is called the *group of holomorphic automorphisms* and denoted by $\mathrm{Aut}(U)$.

Theorem 6.8 *The group* $\mathrm{Aut}(D^2(0,U))$ *consists of all the biholomorphic functions*

$$w = \left(e^{i\theta_1}\frac{z_1 - a_1}{1 - \bar{a}_1 z_1}, e^{i\theta_2}\frac{z_2 - a_2}{1 - \bar{2}_2 z_2}\right) \tag{6.4}$$

and

$$w = \left(e^{i\theta_2}\frac{z_2 - a_1}{1 - \bar{a}_1 z_2}, e^{i\theta_1}\frac{z_1 - a_2}{1 - \bar{a}_2 z_1}\right) \tag{6.5}$$

where $z = (z_1, z_2) \in D^2(0,1)$*,* $a_1, a_2 \in D(0,1)$ *and* $\theta_1, \theta_2 \in [0, 2\pi]$*.*

Proof Let $\varphi(z) \in \mathrm{Aut}(D^2(0,1))$, $\varphi(0) = \alpha = (\alpha_1, \alpha_2)$ and

$$\psi(z) = \left(\frac{z_1 - \alpha_1}{1 - \bar{\alpha}_1 z_1}, \frac{z_2 - \alpha_2}{1 - \bar{\alpha}_2 z_2}\right).$$

Then $g \equiv \psi \circ \varphi \in \mathrm{Aut}(D^2(0,1))$ and $g(0) = 0$. By Theorem 6.7 (Cartan Theorem),

$$g(z) = zA = (z_1, z_2)\begin{pmatrix} a_{11} & a_{12} \\ a_{21} & a_{22} \end{pmatrix} = (a_{11}z_1 + a_{21}z_2, a_{12}z_1 + a_{22}z_2).$$

Since $g \in \mathrm{Aut}(D^2(0,1))$, we have

$$|a_{11}z_1 + a_{21}z_2| < 1, \quad |a_{12}z_1 + a_{22}z_2| < 1$$

for any $(z_1, z_2) \in D^2(0,1)$. It follows that $|a_{ij}| < 1$ $(i, j = 1, 2)$.

Let

$$z^{1,k} = \left(\left(1 - \frac{1}{k}\right)\frac{\bar{a}_{11}}{|a_{11}|}, \quad \left(1 - \frac{1}{k}\right)\frac{\bar{a}_{21}}{|a_{21}|}\right) \in D^2(0,1),$$

$$z^{2,k} = \left(\left(1 - \frac{1}{k}\right)\frac{\bar{a}_{12}}{|a_{12}|}, \quad \left(1 - \frac{1}{k}\right)\frac{\bar{a}_{22}}{|a_{22}|}\right) \in D^2(0,1).$$

Then

$$g(z^{1,k}) = \left(\left(1 - \frac{1}{k}\right)(|a_{11}| + |a_{21}|), *\right) \in D^2(0,1),$$

$$g(z^{2,k}) = \left(*, \left(1 - \frac{1}{k}\right)(|a_{12}| + |a_{22}|)\right) \in D^2(0,1).$$

Hence

$$\left(1-\frac{1}{k}\right)(|a_{11}|+|a_{21}|)<1,$$

$$\left(1-\frac{1}{k}\right)(|a_{12}|+|a_{22}|)<1.$$

Let $k\to\infty$. Then we have

$$|a_{11}|+|a_{21}|\le 1,\quad |a_{12}|+|a_{22}|\le 1. \tag{6.6}$$

On the other hand,

$$\left(1-\frac{1}{k},0\right)\in D^2(0,1),\quad \left(0,1-\frac{1}{k}\right)\in D^2(0,1),$$

$$g\left(1-\frac{1}{k},0\right)=\left(\left(1-\frac{1}{k}\right)a_{11},\quad \left(1-\frac{1}{k}\right)a_{12}\right)\in D^2(0,1),$$

$$g\left(0,1-\frac{1}{k}\right)=\left(\left(1-\frac{1}{k}\right)a_{21},\quad \left(1-\frac{1}{k}\right)a_{22}\right)\in D^2(0,1),$$

and $(1-(1/k),0),(0,1-(1/k))$ approach to $\partial D^2(0,1)$ as $k\to\infty$, so we have

$$(a_{11},a_{12})\in\partial D^2(0,1),\quad (a_{21},a_{22})\in\partial D^2(0,1).$$

Thus

$$\max\{|a_{11}|,|a_{12}|\}=1,\quad \max\{|a_{21}|,|a_{22}|\}=1. \tag{6.7}$$

Only when

$$|a_{11}|=1,\ a_{12}=0,\ a_{21}=0,\ |a_{22}|=1,$$

or

$$|a_{12}|=1,\ a_{11}=0,\ a_{22}=0,\ |a_{21}|=1,$$

(6.6) and (6.7) hold. This is equivalent to that A can only be the following two matrices

$$A=\begin{pmatrix} e^{i\theta_1} & 0\\ 0 & e^{i\theta_2}\end{pmatrix}$$

and

$$A = \begin{pmatrix} 0 & e^{i\theta_1} \\ e^{i\theta_2} & 0 \end{pmatrix}.$$

Hence we have

$$(1) \quad \psi \circ \varphi(z) = z \begin{pmatrix} e^{i\theta_1} & 0 \\ 0 & e^{i\theta_2} \end{pmatrix} = (z_1 e^{i\theta_1}, z_2 e^{i\theta_2}),$$

or

$$(2) \quad \psi \circ \varphi(z) = z \begin{pmatrix} 0 & e^{i\theta_1} \\ e^{i\theta_2} & 0 \end{pmatrix} = (z_2 e^{i\theta_2}, z_2 e^{i\theta_1}).$$

If $\varphi = (\varphi_1, \varphi_2)$, then equation (1) above becomes

$$\left(\frac{\varphi_1 - \alpha_1}{1 - \bar{\alpha}_1 \varphi_1}, \frac{\varphi_2 - \alpha_2}{1 - \bar{\alpha}_2 \varphi_2} \right) = (z_1 e^{i\theta_1},\ z_2 e^{i\theta_2}).$$

That is

$$\frac{\varphi_1 - \alpha_1}{1 - \bar{\alpha}_1 \varphi_1} = z_1 e^{i\theta_1}, \qquad \frac{\varphi_2 - \alpha_2}{1 - \bar{\alpha}_1 \varphi_2} = z_2 e^{i\theta_2}.$$

Solving for φ from the above equations, we get that

$$\varphi = (\varphi_1, \varphi_2) = \left(e^{i\theta_1} \frac{\alpha_1 e^{-i\theta_1} + z_1}{1 + \bar{\alpha}_1 e^{i\theta_1} z_1},\ e^{i\theta_2} \frac{\alpha_2 e^{-i\theta_2} + z_2}{1 + \bar{\alpha}_2 e^{i\theta_2} z_2} \right).$$

This equation is in the form of (6.4). Similarly, from equation (2) above, we can get that φ is in the form of (6.5) and the theorem follows.

Next, we determine the group of holomorphic automorphisms of the unit ball. It can be verified directly that

$$\varphi_a(z_1, z_2) = \left(\frac{z_1 - a}{1 - \bar{a} z_1}, \frac{(1 - |a|^2)^{\frac{1}{2}} z_2}{1 - \bar{a} z_1} \right) \in \mathrm{Aut}(B(0,1)), \tag{6.8}$$

where $a \in \mathbb{C}$ and $|a| < 1$. Since

$$\left| \frac{z_1 - a}{1 - \bar{a} z_1} \right|^2 + \left| \frac{(1 - |a|^2)^{\frac{1}{2}} z_2}{1 - \bar{a} z_1} \right|^2 = \frac{|z_1|^2 - 2\,\mathrm{Re}\,\bar{a} z_1 + |a|^2 + (1 - |a|^2)|z_2|^2}{|1 - \bar{a} z_1|^2},$$

and the right hand side of the above equation is less than or equal to

$$\frac{1 - |a|^2 - 2\,\mathrm{Re}\,\bar{a}z_1 + |a|^2 + |a|^2|z_1|^2}{|1 - \bar{a}z_1|^2} = 1$$

if and only if $(z_1, z_2) \in B(0,1)$. Therefore, (6.8) holds.

It is easy to see that $(\varphi_a)^{-1} = \varphi_{-a}$.

A 2×2 matrix U is called a *unitary matrix* if $U\bar{U}' = I$, where $\bar{U}'$ is the conjugate transpose of U. The map $w = zU$ is called an *unitary rotation* and denoted by $w = U(z)$.

Theorem 6.9 *If $g(z) \in \mathrm{Aut}(B(0,1))$ and $g(0) = 0$, then g is an unitary rotation. In other words, there exists an unitary matrix such that $g(z) = zA$.*

Proof Since $B(0,1)$ is a circular region, by Theorem 6.7 (Cartan Theorem), $g(z) \equiv zA$ and g maps unit vectors to unit vectors. If

$$A = \begin{pmatrix} a_{11} & a_{12} \\ a_{21} & a_{22} \end{pmatrix}$$

and (α, β) is a unit vector, then

$$(\alpha, \beta) \begin{pmatrix} a_{11} & a_{12} \\ a_{21} & a_{22} \end{pmatrix} = (\gamma, \delta)$$

is also a unit vector and

$$\gamma = a_{11}\alpha + a_{21}\beta, \quad \delta = a_{21}\alpha + a_{22}\beta.$$

Thus

$$|a_{11}\alpha + a_{21}\beta|^2 + |a_{12}\alpha + a_{22}\beta|^2 = 1. \tag{6.9}$$

Let $\alpha = 1, \beta = 0$ and $\alpha = 0, \beta = 1$. Then

$$|a_{11}|^2 + |a_{12}|^2 = 1, \quad |a_{21}|^2 + |a_{22}|^2 = 1 \tag{6.10}$$

respectively.

Substitute (6.10) in (6.9), we have

$$\mathrm{Re}((a_{11}\bar{a}_{21} + a_{12}\bar{a}_{22})\alpha\bar{\beta}) = 0.$$

Let $\alpha = 1/\sqrt{2}, \beta = 1/\sqrt{2}$ and $\alpha = i/\sqrt{2}, \beta = 1/\sqrt{2}$. Then

$$\mathrm{Re}(a_{11}\bar{a}_{21} + a_{12}\bar{a}_{22}) = 0,$$

$$\mathrm{Im}(a_{11}\bar{a}_{21} + a_{12}\bar{a}_{22}) = 0.$$

Therefore

$$a_{11}\bar{a}_{21} + a_{12}\bar{a}_{22} = 0. \tag{6.11}$$

By (6.10) and (6.11), the conclusion that A is an unitary matrix follows.

Theorem 6.10 *Every element of* $\mathrm{Aut}(B(0,1))$ *is a composition of at most two unitary rotations and one* φ_a. *In other words,* $\mathrm{Aut}(B(0,1))$ *consists of unitary rotations,* φ_a *and their compositions.*

Proof If $f \in \mathrm{Aut}(B(0,1))$ and $f(0) = \alpha$, then there exists an unitary matrix U such that $\alpha U = (|\alpha|, 0)$. Let $g(z) = \varphi_{|a|} \circ U \circ f(z)$, where $\varphi_{|a|}$ is defined by (6.8) which maps $(|\alpha|, 0)$ to 0. Then $g(z) \in \mathrm{Aut}(B(0,1))$ and

$$g(0) = \varphi_{|\alpha|} \circ U \circ f(0) = \varphi_{|a|} \circ U \circ \alpha = 0.$$

By Theorem 6.9, we have

$$g(z) = zV = V(z),$$

where V is an unitary matrix. Therefore,

$$f(z) = U^{-1} \circ \varphi_{-|\alpha|} \circ V(z).$$

This proves Theorem 6.10.

6.4 Poincaré Theorem

Now we are ready to prove the important Poincaré Theorem.

Theorem 6.11 ***(Poincaré Theorem)*** *There are no biholomorphic functions which map* $D^2(0,1)$ *onto* $B(0,1)$.

Proof Suppose the contrary. If there exists such a biholomorphic function φ which maps $D^2(0,1)$ to $B(0,1)$ and $\varphi(0) = \alpha$, then $\Phi = \varphi_\alpha \circ \varphi$ is also a biholomorphic function which maps $D^2(0,1)$ to $B(0,1)$ and $\Phi(0) = \varphi_\alpha \circ \varphi(0) = 0$, where φ_α is defined by (6.8). If $h \in \mathrm{Aut}(D^2(0,1))$, then

$$h \to \Phi \circ h \circ \Phi^{-1} \in \mathrm{Aut}(B(0,1)) \tag{6.12}$$

is an isomorphism between the two groups. Let $(\mathrm{Aut}(D^2(0,1)))_0$ and $(\mathrm{Aut}(B(0,1)))_0$ be the branches of $\mathrm{Aut}(D^2(0,1))$ and $\mathrm{Aut}(B(0,1))$ respectively which contain the unit elements. Then (6.12) is also an isomorphism between $(\mathrm{Aut}(D^2(0,1)))_0$ and $(\mathrm{Aut}(B(0,1)))_0$. Especially, (6.12) is an isomorphism between $\mathrm{Aut}_0(D^2(0,1))$ and $\mathrm{Aut}_0(B(0,1))$, where these two subgroups consist of the automorphisms that map the origin to the origin.

According to Theorem 6.8, $\mathrm{Aut}_0(D^2(0,1))$ consists of all the biholomorphic functions with the form

$$w = (e^{i\theta_1}z_1,\, e^{i\theta_2}z_2) = (z_1, z_2)\begin{pmatrix} e^{i\theta_1} & 0 \\ 0 & e^{i\theta_2} \end{pmatrix}$$

where θ_1 and θ_2 are real numbers. In other words, this group consists of

$$\left\{\begin{pmatrix} e^{i\theta_1} & 0 \\ 0 & e^{i\theta_2} \end{pmatrix}\right\}.$$

By Theorem 6.10, $\mathrm{Aut}_0(B(0,1))$ consists of all the biholomorphic maps $w = zVU^{-1}$, where U and V are unitary matrices. Since the inverse and the product of unitary matrices are also unitary matrices, $\mathrm{Aut}_0(B(0,1))$ consists of all $w = zX$, where X is an unitary matrix. In other words, this group consists of all unitary matrices and it is called the *unitary group*.

If there exists a biholomorphic function which maps $D^2(0,1)$ to $B(0,1)$, then (6.12) establishes a group isomorphism from $\mathrm{Aut}_0(D^2(0,1))$ to $\mathrm{Aut}_0(B(0,1))$. This implies that the group

$$\left\{\begin{pmatrix} e^{i\theta_1} & 0 \\ 0 & e^{i\theta_2} \end{pmatrix}\right\}$$

is isomorphic to the unitary group. This is impossible since the group

$$\left\{\begin{pmatrix} e^{i\theta_1} & 0 \\ 0 & e^{i\theta_2} \end{pmatrix}\right\}$$

is an abelian group, but the unitary group is not. This is a contradiction. Therefore, such a biholomorphic map does not exists and the Poincaré theorem follows.

According to the Riemann Mapping Theorem in Chapter 4, for any simply connected region Ω whose boundary contains at least two points, there exists an univalent holomorphic function which maps U to the unit disc. In other words, any two topologically equivalent regions must be holomorphically equivalent. By the Poincaré Theorem, the above assertion is not true in $\mathbb{C}^n$ with $n \geq 2$. That two regions are topologically equivalent does not necessarily imply that they are holomorphically equivalent. Hence, it brings up the classification problem of regions in $\mathbb{C}^n$. If two regions are topologically equivalent, under what condition are they also holomorphically equivalent? This is still an unsolved problem. What we know is that almost all regions that are topologically equivalent are not holomorphically equivalent.

On the other hand, we can see that, the Riemann Mapping Theorem in Chapter 4 holds only in one dimension and it is a deep result in complex analysis. From here, we can get several important theorems which are also true only in the one dimensional case.

6.5 Hartogs Theorem

In the complex analysis of one variable, if Ω is a region in $\mathbb{C}$ and $a \in \mathbb{C} \setminus \Omega$, then there exists a holomorphic function f in Ω such that this function does not have the analytic continuation at point a. For instance, $f(z) = 1/(z-a)$ is such a function. In the function theory of several complex variables, the above conclusion is not true. This is called the *Hartogs phenomenon.*

Theorem 6.12 ***(Hartogs Theorem)*** *Suppose $\Omega \subseteq \mathbb{C}^n$ $(n \geq 2)$ is a region, K is a compact subset of Ω and $\Omega \setminus K$ is connected. If f is a holomorphic function on $\Omega \setminus K$, then there exists a holomorphic function F on Ω such that F is equal to f on $\Omega \setminus K$. In other words, if a function is holomorphic on $\Omega \setminus K$, then it has an analytic continuation on Ω.*

We only prove some special cases of this theorem here.

Let R be a region of $\mathbb{C}^2$. If $z = (z_1, z_2) \in R$ implies that $(e^{i\theta_1} z_1, e^{i\theta_2} z_2) \in R$ for any real number θ_1 and θ_2, then we say that R is a *Reinhardt domain.*

Theorem 6.13 *Let R be a Reinhardt domain in $\mathbb{C}^2$ and $f(z)$ be a holomorphic function. Then f has the Laurent expansion on R*

$$\sum_{j,k=-\infty}^{\infty} a_{jk} z_1^j z_2^k. \tag{6.13}$$

This series converges uniformly to f on any compact subset of R and such an expansion is unique.

Proof First, we prove the uniqueness. Let $w = (w_1, w_2) \in R$ and $w_1 \neq 0, w_2 \neq 0$. Since (6.13) converges uniformly on every compact subset of R, if

$z_1 = w_1 e^{i\theta_1}$ and $z_2 = w_2 e^{i\theta_2}$, then $(z_1, z_2) \in R$ and the subset that consists of all (z_1, z_2) is a compact subset of R for $-\pi \leq \theta_1 \leq \pi$ and $-\pi \leq \theta_2 \leq \pi$. Thus

$$a_{jk} = \frac{w_1^{-j} w_2^{-k}}{(2\pi)^2} \int_{-\pi}^{\pi} \int_{-\pi}^{\pi} f(w_1 e^{i\theta_1}, w_2 e^{i\theta_2}) e^{-i(j\theta_1 + k\theta_2)} \, d\theta_1 d\theta_2.$$

This implies that all the a_{jk}s are uniquely determined by f.

Next, we prove the existence of (6.13).

It is easy to see that, if $f(z)$ is holomorphic on $\Omega = \{z \in \mathbb{C}^2 | \ r_1 < |z_1| < R_1, r_2 < |z_2| < R_2\}$, then perform the one dimensional Laurent expansion twice with respect to z_1 and z_2, we get the Laurent series of $f(z)$ on Ω

$$\sum_{j,k=-\infty}^{\infty} b_{jk} z_1^j z_2^k,$$

and it converges uniformly on any compact subset of Ω.

If $w = (w_1, w_2) \in R$, then we can find a sufficiently small ε such that

$$\begin{aligned} \Omega(w, \varepsilon) =& \{z \in \mathbb{C}^2 | \ |w_1| - \varepsilon < |z_1| < |w_1| + \varepsilon, \\ & |w_2| - \varepsilon < |z_2| < |w_2| + \varepsilon\} \subseteq R. \end{aligned}$$

Thus $f(z)$ has a Laurent expansion on $\Omega(w, \varepsilon)$

$$f(z) = \sum_{j,k=-\infty}^{\infty} a_{jk}(w) z_1^j z_2^k, \quad z \in \Omega,$$

and this series converges uniformly to f on a neighborhood of w.

If $w' \in \Omega(w, \varepsilon)$ and $f(z)$ has a Laurent expansion at point w'

$$\sum_{j,k=-\infty}^{\infty} a_{jk}(w') z_1^j z_2^k,$$

then by the uniqueness of the Laurent expansion we have $a_{jk}(w') = a_{jk}(w)$. In other words, $a_{jk}(w)$ is locally constant in R. Since R is connected, $a_{jk}(w) = a_{jk}$ is a constant which does not depend on w. Thus $f(z)$ has a Laurent expansion in R

$$\sum_{j,k=-\infty}^{\infty} a_{jk} z_1^j z_2^k,$$

and it converges uniformly on a neighborhood of any point $z \in R$. Hence it converges uniformly on every compact subset of R. This proves the theorem.

Theorem 6.13 implies that

Theorem 6.14 *Let R be a Reinhardt domain in $\mathbb{C}^2$ that contains points with the first coordinate zero and points with the second coordinate zero. That is, R contains a $(0, z_2)$ and a $(z_1, 0)$. Then any holomorphic function on R has the expansion*

$$f(z) = \sum_{j,k \geq 0} a_{jk} z_1^j z_2^k, \tag{6.14}$$

and this series converges uniformly on any compact subset of R.

Proof By Theorem 6.13, $f(z)$ has the expansion (6.13). Since R contains the point $(0, z_2)$, all the a_{jk}s are zero in (6.13) when $j < 0$. Otherwise, (6.13) would not converge uniformly in the neighborhoods of $(0, z_2)$. Similarly, since R contains the point $(z_1, 0)$, all the a_{jk}s are zero in (6.13) when $k < 0$. Thus the theorem follows.

Theorem 6.15 ***(Hartogs Theorem on Reinhardt Domains)*** *Let R be a Reinhardt domain in $\mathbb{C}^2$ that contains points with the first coordinate zero and points with the second coordinate zero. That is, R contains a $(0, z_2)$ and a $(0, z_1)$. Then any holomorphic function $f(z)$ on R can be analytically continued to*

$$R' = \{(\rho_1 z_1, \rho_2 z_2) \in \mathbb{C}^2 | \; 0 \leq \rho_1 \leq 1, 0 \leq \rho_2 \leq 1, \; (z_1, z_2) \in R\}.$$

In other words, there exists a holomorphic function F on R' such that $f = F$ for $z \in R$.

The proof of this theorem is obvious. Since $f(z)$ can be expanded as (6.14) in R, for any $z \in R$, the series (6.14) converges uniformly in the neighborhoods of z. Since $(\rho_1 z_1, \rho_2 z_2) \in R'$, the series

$$\sum_{j,k \geq 0} a_{jk} \rho_1^j \rho_2^k z_1^j z_2^k$$

is convergent. Hence, (6.14) converges uniformly on the neighborhoods of $(\rho_1 z_1, \rho_2 z_2)$. If (6.14) converges to a function F, then F is just what we wanted.

The following is an example of Hartogs phenomenon.

Example 6.1 Let

$$B_r(0,1) = \{z = (z_1, z_2) \in \mathbb{C}^2 |\ r < |z_1|^2 + |z_2|^2 < 1\}.$$

Then $B_r(0,1)$ is a Reinhardt domain, where $0 < r < 1$. Thus, if $f(z)$ is holomorphic on $B_r(0,1)$, then it can be analytically continued to

$$B(0,1) = \{z = (z_1, z_2) \in \mathbb{C}^2 |\ |z_1|^2 + |z_2|^2 < 1\}.$$

That is, it can be analytically continued to the unit ball.

We have discussed the Hartogs Theorem on the Reinhardt domain. The kind of regions without Hartogs phenomenon is called *domains of holomorphy*. The study of domains of holomorphy is one of the main topic in the research of the function theory of several complex variables in the last century. The reader can refer to books about the function theory of several complex variables such as S. G. Krantz [2] and R. Narasimhan [2].

The discussion of this chapter is in $\mathbb{C}^2$. It is easy to see that these results can be generalized to $\mathbb{C}^n$ without any difficulties.

Chapter 7

Elliptic Functions

7.1 The Concept of Elliptic Functions

In this and the next chapters, we introduce two classes of important complex functions, elliptic functions and the Riemann ζ-function. Developments of many branches of modern mathematics are closely related to these functions. Extensive literature exists about the many beautiful results of these functions. We give a brief introduction here to familiarize the readers with these functions.

The theory of elliptic functions is a fancy chapter in complex analysis. it originated from some age-old problems such as finding the length of an arc on an ellipse, finding a conformal mapping from half planes to rectangles using Schwarz-Christoffel formula (Section 4.5). But the most natural way to introduce elliptic functions is by periodic functions.

In calculus, a periodic function can be expressed as a Fourier expansion, namely by the simplest periodic functions $\sin x$, $\cos x$, $\cdots$, $\sin nx$, $\cos nx$, $\cdots$. Is there any similar expression for a periodic function on $\mathbb{C}$? The function e^z is the simplest singly periodic holomorphic function with period $2\pi i$, and the function $e^{2\pi iz/\omega}$ is a singly periodic (with period ω) holomorphic function. If $f(z)$ is a singly periodic (with period ω) meromorphic function on $\mathbb{C}$, then it can be expressed in terms of $e^{2\pi iz/\omega}$ as the complex Fourier development of $f(z)$.

$$f(z) = \sum_{n=-\infty}^{\infty} c_n e^{\frac{2\pi inz}{\omega}}$$

where

$$c_n = \frac{1}{\omega}\int_a^{a+\omega} f(z)e^{\frac{-2\pi inz}{\omega}}\,dz, \quad a \in \mathbb{C}.$$

We leave the verification of this conclusion as an exercise to the reader.

By an elliptic function, we mean a meromorphic function $f(z)$ on $\mathbb{C}$ with two periods ω_1, ω_2 such that $\mathrm{Im}(\omega_2/\omega_1) > 0$. (The fact that ω_1 and ω_2 are not collinear leads to the condition $\mathrm{Im}(\omega_2/\omega_1) \neq 0$. We always assume that $\mathrm{Im}(\omega_2/\omega_1) > 0$, otherwise consider ω_1, $-\omega_2$ or $-\omega_1$, ω_2.)

Similar to the case of singly periodic functions, by a simple verification, we can see that a period ω of a doubly periodic function can be uniquely expressed by

$$\omega = m\omega_1 + n\omega_2,$$

where m and n are integers.

All points ω in the set $M = \{m\omega_1 + n\omega_2| \quad m, n \in \mathbb{Z}\}$ form geometrically a lattice L in the complex plane. The set M is a discrete additive subgroup of $\mathbb{C}$, or in other words, a free $\mathbb{Z}$-module of rank 2 with ω_1, ω_2 as a $\mathbb{Z}$-basis.

For any $\alpha \in \mathbb{C}$, the set of complex numbers

$$P = \{\alpha + t_1\omega_1 + t_2\omega_2| \quad t_1, t_2 \in [0, 1)\}$$

is called a *fundamental parallelogram* for L. Any element of $\mathbb{C}/L$ has a

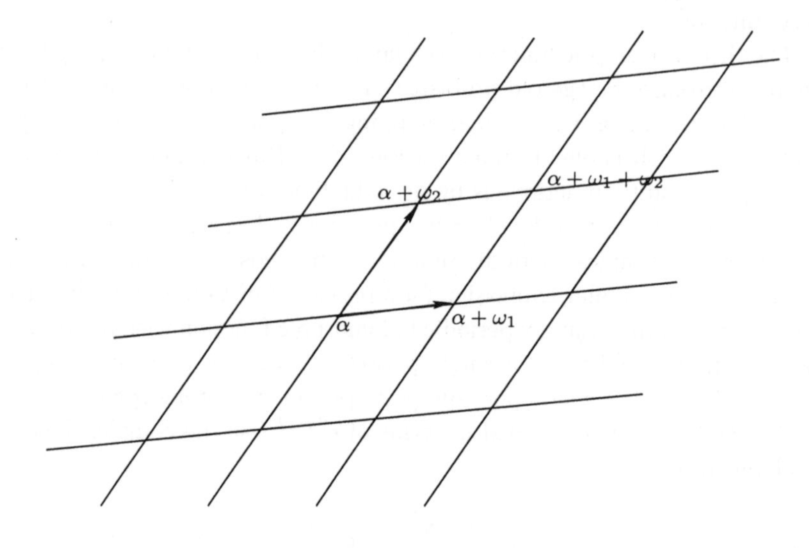

Fig. 9

unique representative in P with respect to a given point. We can obtain properties of an elliptic function $f(z)$ on the whole plane from properties of $f(z)$ on a fundamental parallelogram since $f(z)$ is a doubly periodic function.

It is clear that the set of elliptic functions with the usual arithmetic operations of functions is a field which is closed under differentiation. We leave the verification to the reader.

Since $\mathbb{C}/L$ is homeomorphic to a torus, it is compact. Therefore, an elliptic function without poles must be a constant function by Liouville theorem.

Several properties about zeros and poles of elliptic functions are based on the next theorem.

Theorem 7.1 *The number of poles and zeros of an elliptic function $f(z)$ in $\mathbb{C}/L$ is finite.*

Proof It suffices to show that $f(z)$ has only finite number of poles and zeros in $\bar{P}$, the closure of P. Suppose the contrary, $f(z)$ has an infinite number of poles in $\bar{P}$. The compactness of $\bar{P}$ implies the Bolzano-Weierstrass property (Theorem 1.11 in Section 1.3) so that there is a limit point of the infinite set of poles. This contradicts with the assumption that $f(z)$ is meromorphic and it can only have isolated singularities. Consider $1/f(z)$ and though the same discussion we see that $f(z)$ has only finite number of zeros in $\bar{P}$. The theorem is proved.

By this theorem, we can always choose an $\alpha \in \mathbb{C}$ such that $f(z)$ has no zeros nor poles on ∂P.

Theorem 7.2 *The sum of the residues of an elliptic function $f(z)$, at the poles of $f(z)$ on $\mathbb{C}/L$, is zero.*

Proof Choose a fundamental parallelogram P with vertices α, $\alpha + \omega_1$, $\alpha + \omega_1 + \omega_2$, $\alpha + \omega_2$ (as in Fig. 9) such that $f(z)$ has no poles on ∂P. By the residue theorem (Theorem 3.11 in Section 3.5) and the periodicity of $f(z)$, the sum of residues of $f(z)$ in P is

$$\begin{aligned}
\frac{1}{2\pi i}\int_{\partial P} f(z)\,dz &= \frac{1}{2\pi i}\left(\int_{\alpha}^{\alpha+\omega_1} f(z)\,dz + \int_{\alpha+\omega_1}^{\alpha+\omega_1+\omega_2} f(z)\,dz\right. \\
&\quad \left. + \int_{\alpha+\omega_1+\omega_2}^{\alpha+\omega_2} f(z)\,dz + \int_{\alpha+\omega_2}^{\alpha} f(z)\,dz\right) \\
&= \frac{1}{2\pi i}\left(\int_{\alpha}^{\alpha+\omega_1} f(z)\,dz + \int_{\alpha}^{\alpha+\omega_2} f(z)\,dz\right. \\
&\quad \left. + \int_{\alpha+\omega_1}^{\alpha} f(z)\,dz + \int_{\alpha+\omega_2}^{\alpha} f(z)\,dz\right) = 0
\end{aligned}$$

We need the following preparations to continue our discussion about elliptic functions.

Let Γ be a piecewise smooth simple closed curve on $\mathbb{C}$, $f(z)$ be a function holomorphic inside and on Γ except that it may have finitely many poles inside Γ. Also assume that $f(z)$ does not vanished on Γ. Denote $a_1, a_2, \cdots, a_k$ as zeros and $b_1, b_2, \cdots, b_m$ as poles of $f(z)$ inside Γ with multiplicities $\alpha_1, \alpha_2, \cdots, \alpha_k$ and $\beta_1, \beta_2, \cdots, \beta_m$ respectively. Then for any function $\varphi(z)$ holomorphic inside and on Γ, we have

$$\frac{1}{2\pi i}\int_\Gamma \varphi(z)\frac{f'(z)}{f(z)}\,dz = \sum_{i=1}^{k}\alpha_i\varphi(a_i) - \sum_{j=1}^{m}\beta_j\varphi(b_j).$$

Let $\varphi(z) = 1$. Then the logarithmic residue of $f(z)$ with respect to Γ is equal to the difference between the number of zeros inside Γ and the number of poles of $f(z)$ inside Γ, the zeros and poles being counted according to their multiplicities. That is,

$$\frac{1}{2\pi i}\int_\Gamma \frac{f'(z)}{f(z)}\,dz = \sum_{i=1}^{k}\alpha_i - \sum_{j=1}^{m}\beta_j,$$

where $\sum_{i=1}^{k}\alpha_i = N$ is the number of zeros of $f(z)$ inside Γ and $\sum_{j=1}^{m}\beta_j = S$ is the number of poles of $f(z)$ inside Γ.

Applying this result to the function $f(z) - \alpha$ with an arbitrary complex number α, and with the same assumption, we have that the value of the integral

$$\frac{1}{2\pi i}\int_\Gamma \frac{f'(z)}{f(z)-\alpha}\,dz$$

is equal to the difference between the number of roots of the equation $f(z) = \alpha$ inside Γ and the number of poles of $f(z)$ inside Γ, each counted with their multiplicities.

Let $\varphi(z) = z$. Then we have

$$\frac{1}{2\pi i}\int_\Gamma z\frac{f'(z)}{f(z)}\,dz = \sum_{i=1}^{k}\alpha_i a_i - \sum_{j=1}^{m}\beta_j b_j.$$

This equation gives the difference between the sum of zeros and the sum of poles of $f(z)$ inside Γ, each counted with their multiplicities.

These results can be proved by the residue theorem. We suggest the reader to verify them as exercises before checking other textbooks.

Theorem 7.3 *The number of zeros of a non-constant elliptic function $f(z)$ in a fundamental parallelogram P is equal to the number of poles of $f(z)$ in P. The zeros and poles being counted according to their multiplicities.*

Proof Choose a P such that $f(z)$ has no poles nor zeros on ∂P. Since the field of elliptic functions is closed under differentiation, the function $f'(z)/f(z)$ is elliptic. The sum of its residues in P which is calculated by the integral

$$\int_{\partial P} \frac{f'(z)}{f(z)}\, dz$$

is zero. But the value of this integral is also the difference between the number of zeros and the number of poles of $f(z)$ in P, each counted with their multiplicities, by the results we just stated above. The theorem follows.

Theorem 7.4 *A non-constant elliptic function $f(z)$ assumes in a fundamental parallelogram P every complex value (finite or infinite) the same number of times.*

Proof Let α be an arbitrary complex number. Then the function

$$F(z) = \frac{f'(z)}{f(z) - \alpha}$$

is elliptic. Choosing a P such that $F(z)$ has no poles nor zeros on ∂P, then

$$\frac{1}{2\pi i} \int_{\partial P} F(z)\, dz = \frac{1}{2\pi i} \int_{\partial P} \frac{f'(z)}{f(z) - \alpha}\, dz = 0.$$

From the results we stated above, this integral is equal to the difference between the number of zeros and the number of poles of $f(z) - \alpha$ in P. Since the value of the integral is zero, it follows that the number of roots of the equation $f(z) = \alpha$ inside P is equal to the number of poles of $f(z)$ in P. Hence the theorem is proved.

The concept of the order of an elliptic function follows from this theorem.

Definition 7.1 An elliptic function $f(z)$ has order s if it assumes in a fundamental parallelogram every complex value exactly s times, each value being counted according to its multiplicity. Equivalently, as defined in some other textbooks, the *order of an elliptic function* is the number of its poles in a fundamental parallelogram, each pole being counted according to its multiplicity.

From Theorem 7.2, the order of a non-constant elliptic function is greater than or equal to two.

Theorem 7.5 *The difference between the sum of zeros and the sum of poles of a non-constant elliptic function $f(z)$ on a fundamental parallelogram P is equal to a period of $f(z)$, the zeros and poles being counted according to their multiplicity.*

Proof Choose P as in Fig. 9 such that $f(z)$ has no poles nor zeros on ∂P. Let $a_1, \cdots, a_n$ be the zeros and $b_1, \cdots, b_n$ be the poles of $f(z)$ in P, each appearing as many times as its multiplicity. By the results stated above and Theorem 7.2, we have

$$\begin{aligned}\sum_{k=1}^{n} a_k - \sum_{k=1}^{n} b_k &= \frac{1}{2\pi i}\int_{\partial P} z\frac{f'(z)}{f(z)}\,dz \\ &= \frac{1}{2\pi i}\left(\int_{\alpha}^{\alpha+\omega_1} z\frac{f'(z)}{f(z)}\,dz + \int_{\alpha+\omega_1}^{\alpha+\omega_1+\omega_2} z\frac{f'(z)}{f(z)}\,dz\right. \\ &\quad \left. + \int_{\alpha+\omega_1+\omega_2}^{\alpha+\omega_2} z\frac{f'(z)}{f(z)}\,dz + \int_{\alpha+\omega_2}^{\alpha} z\frac{f'(z)}{f(z)}\,dz\right). \qquad (7.1)\end{aligned}$$

Let $z = \zeta + \omega_2$. Then the third integral inside the parentheses is

$$\begin{aligned}\int_{\alpha+\omega_1+\omega_2}^{\alpha+\omega_2} z\frac{f'(z)}{f(z)}\,dz &= -\int_{\alpha+\omega_2}^{\alpha+\omega_1+\omega_2} z\frac{f'(z)}{f(z)}\,dz \\ &= -\int_{\alpha}^{\alpha+\omega_1} (\zeta+\omega_2)\frac{f'(\zeta+\omega_2)}{f(\zeta+\omega_2)}\,d\zeta \\ &= -\int_{\alpha}^{\alpha+\omega_1} (\zeta+\omega_2)\frac{f'(\zeta)}{f(\zeta)}\,d\zeta \\ &= -\int_{\alpha}^{\alpha+\omega_1} \zeta\frac{f'(\zeta)}{f(\zeta)}\,d\zeta - \omega_2\int_{\alpha}^{\alpha+\omega_1} \frac{f'(\zeta)}{f(\zeta)}\,d\zeta.\end{aligned}$$

It follows that the sum of the first and the third integral in the parentheses of (7.1) is equal to

$$-\omega_2\int_{\alpha}^{\alpha+\omega_1} \frac{df(\zeta)}{f(\zeta)} = -\omega_2[\ln f(\alpha+\omega_1) - \ln f(\alpha)].$$

By the Argument Principle (Theorem 2.14 in Section 2.4) and the periodicity of $f(z)$, the expression in the bracket equals $2\pi i l$ for an integer l. Similarly, the sum of the second and the forth integral in the parentheses of (7.1) is equal to $2\pi i s\omega_1$ for an integer s. The theorem is proved.

7.2 The Weierstrass Theory

By the previous discussion, we know that there exists no elliptic functions with order one. So we start our study from elliptic functions with order two. Such functions have either a double pole or two simple poles in a fundamental parallelogram. These are the different starting points of the Weierstrass theory and the Jacobi theory of elliptic functions. More precisely, Weierstrass constructed a function with a double pole and developed a systematic theory from it, while Jacobi theory comes from the study of the case of two simple poles.

In further studies of this subject, the reader will find that Jacobi theory is more often used in solving practical problems and Weierstrass theory is more convenient in theoretical developments.

Let $\omega = m\omega_1 + n\omega_2$ where $m, n \in \mathbb{Z}$ and $(m, n) \neq (0, 0)$. Then the series

$$\sum_{\omega \neq 0} \frac{1}{|\omega|^\alpha} \tag{7.2}$$

converges when $\alpha > 2$.

In fact, let P_n, $(n = 1, 2, \cdots)$, be a parallelogram with center $z = 0$ and one of its vertices $n(\omega_1 + \omega_2)$ (see Fig. 10). Then there are $8n$ periods on ∂P_n. Let $\delta > 0$ be the distance from $z = 0$ to ∂P_1 (Fig. 10). Then the

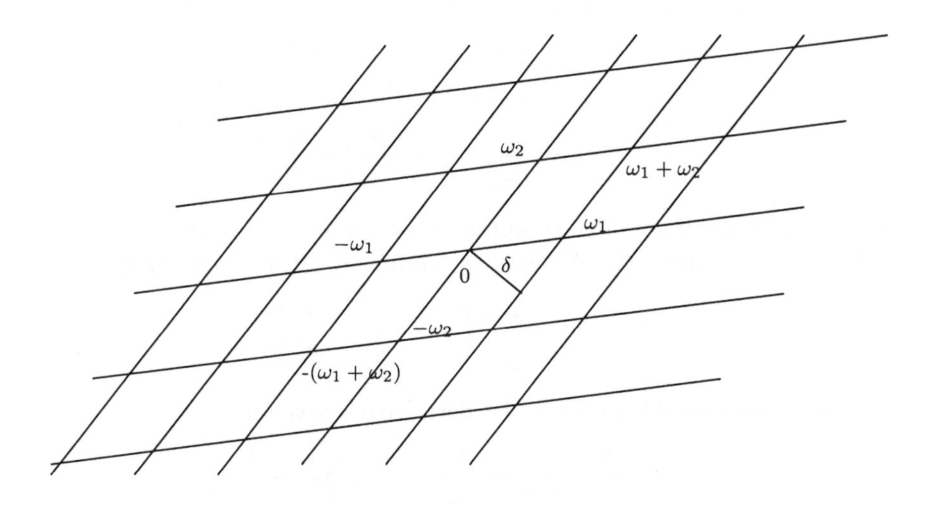

Fig. 10

distance from $z = 0$ to ∂P_n is $n\delta$. Thus

$$\sum_{\omega \in \partial P_n} \frac{1}{|\omega|^\alpha} \le \frac{8n}{(n\delta)^\alpha} = \frac{8}{\delta^\alpha} \frac{1}{n^{\alpha-1}}.$$

Therefore (7.2) converges when $\alpha > 2$ since

$$\sum_{n=1}^{\infty} \frac{1}{n^{\alpha-1}}$$

converges when $\alpha > 2$.

From this result, we can deduce that the series

$$\sum_{\omega \neq 0} \left\{ \frac{1}{(z-\omega)^2} - \frac{1}{\omega^2} \right\} \tag{7.3}$$

converges uniformly on any compact set that does not intersect M. Since the convergence will not be affected by the omission of a sufficient number of initial terms in (7.2), without loss of generosity, we can assume that $|\omega| > 2R$ for an arbitrary $R > 0$. Thus, if $|z| \le R$, we have

$$\frac{|z|}{|\omega|} < \frac{1}{2}, \quad \left|2 - \frac{z}{\omega}\right| \le 2 + \frac{1}{2}, \quad \left|1 - \frac{z}{\omega}\right|^2 \ge \left(1 - \frac{1}{2}\right)^2.$$

It follows that

$$\begin{aligned} \left| \frac{1}{(z-\omega)^2} - \frac{1}{\omega^2} \right| &= \left| \frac{z(2\omega - z)}{\omega^2 (z-\omega)^2} \right| \\ &= \left| \frac{\omega z \left(2 - \frac{z}{\omega}\right)}{\omega^4 \left(1 - \frac{z}{\omega}\right)^2} \right| \le \frac{|\omega z|(2 + \frac{1}{2})}{|\omega|^4 \left(1 - \frac{1}{2}\right)^2} \\ &= \frac{|z|}{|\omega|^3} \cdot \frac{\frac{5}{2}}{\frac{1}{4}} = \frac{10|z|}{|\omega|^3} \le \frac{10R}{|\omega|^3}. \end{aligned}$$

So by the convergence of (7.2), our assertion about (7.3) follows.

Now we introduce the Weierstrass $\wp$-function which is defined by

$$\wp(z) = \frac{1}{z^2} + \sum_{\omega \neq 0} \left\{ \frac{1}{(z-\omega)^2} - \frac{1}{\omega^2} \right\}.$$

This is a meromorphic function. Its derivative is given by

$$\wp'(z) = -\frac{2}{z^3} - \sum_{\omega \neq 0} \frac{2}{(z-\omega)^3} = -2 \sum_{\omega} \frac{1}{(z-\omega)^3}.$$

It converges absolutely for $z \notin L$. Therefore, it is also a meromorphic function. Since the series for $\wp'(z+\omega_1)$ is just a rearrangement of the series for $\wp'(z)$, we have

$$\wp'(z+\omega_1) = \wp'(z).$$

Similarly,

$$\wp'(z+\omega_2) = \wp'(z).$$

Hence $\wp'(z)$ is an elliptic function. Integrating the above equations, we get

$$\wp(z+\omega_1) = \wp(z) + c_1,$$

$$\wp(z+\omega_2) = \wp(z) + c_2,$$

where c_1 and c_2 are constants. Setting $z = -\omega_1/2$ (not a pole of $\wp$), we get

$$\wp\left(\frac{\omega_1}{2}\right) = \wp\left(-\frac{\omega_1}{2}\right) + c_1.$$

Since $\wp(z)$ is even, $c_1 = 0$. Similarly, $c_2 = 0$. So $\wp(z)$ is an elliptic function of order two and periods ω_1, ω_2. The points $z = \omega = m\omega_1 + n\omega_2$; $m, n = 0, \pm 1, \pm 2, \cdots$ are double poles of $\wp(z)$. These points are equivalent in $\mathbb{C}/L$. The fundamental parallelogram is homeomorphic to a torus, the points 0, ω_1, ω_2 and $\omega_1 + \omega_2$ on P, a fundamental parallelogram with respect to 0, become one point on the torus. Therefore, the function $\wp(z)$ is of the type we want to find.

The Laurent expansion of $\wp(z)$ at $z = 0$ is

$$\begin{aligned}
\wp(z) &= \frac{1}{z^2} + \sum_{\omega \neq 0} \left[\frac{1}{\omega^2}\left(1 + \frac{z}{\omega} + \left(\frac{z}{\omega}\right)^2 + \cdots\right)^2 - \frac{1}{\omega^2}\right] \\
&= \frac{1}{z^2} + \sum_{\omega \neq 0} \left[\frac{1}{\omega^2}\left(1 + 2\frac{z}{\omega} + 3\left(\frac{z}{\omega}\right)^2 + \cdots\right) - \frac{1}{\omega^2}\right] \\
&= \frac{1}{z^2} + \sum_{\omega \neq 0} \sum_{m=1}^{\infty} (m+1)\left(\frac{z}{\omega}\right)^m \frac{1}{\omega^2} \\
&= \frac{1}{z^2} + \sum_{m=1}^{\infty} c_m z^m
\end{aligned}$$

where $c_m = \sum_{\omega \neq 0}(m+1)/\omega^{m+2}$. Clearly, $c_m = 0$ if m is odd. Thus

$$\wp(z) = \frac{1}{z^2} + \sum_{n=1}^{\infty} c_{2n} z^{2n},$$

where

$$c_{2n} = \sum_{\omega \neq 0} \frac{2n+1}{\omega^{2n+2}} = (2n+1) \sum_{\omega \neq 0} \frac{1}{\omega^{2n+2}},$$

and

$$\wp'(z) = -\frac{2}{z^3} + \sum_{n=1}^{\infty} 2n c_{2n} z^{2n-1}.$$

Adopting the notation

$$b_m = \sum_{\omega \neq 0} \frac{1}{\omega^m}$$

we get

$$\wp(z) = \frac{1}{z^2} + 3b_4 z^2 + 5b_6 z^4 + \cdots, \tag{7.4}$$

and

$$\wp'(z) = -\frac{2}{z^3} + 6b_4 z + 20 b_6 z^3 + \cdots. \tag{7.5}$$

Moreover, by expanding the expressions of $\wp'^2(z)$ and $\wp^3(z)$ from (7.4) and (7.5), we obtain a differential equation satisfied by $\wp(z)$

$$\wp'^2(z) = 4\wp^3(z) - g_2\wp(z) - g_3 \tag{7.6}$$

where $g_2 = 60b_4$ and $g_3 = 140b_6$. We omit the proof of this assertion here and the reader can verify it as an exercise.

Let $u = \wp(z)$. Solving for z in (7.6), we get

$$z = \int_{\infty}^{u} \frac{du}{\sqrt{4u^3 - g_2 u - g_3}} \tag{7.7}$$

which shows that $\wp(z)$ is the inverse of an elliptic integral. Obviously, the polynomial under the square root cannot have any root with multiplicity greater than one, otherwise (7.7) can be expressed as an elementary function. Conversely, it can be shown that for any given g_2 and g_3 such that the polynomial under the square root has distinct zeros, the inverse function of integral (7.7) is a Weierstrass $\wp$-function.

Let e_1, e_2 and e_3 be the three zeros of the polynomial. Then (7.6) becomes

$$\wp'^2(z) = 4(\wp(z) - e_1)(\wp(z) - e_2)(\wp(z) - e_3). \tag{7.8}$$

Since $\wp(\omega_1 - z) = \wp(-z) = \wp(z)$, we have $-\wp'(\omega_1 - z) = \wp'(z)$. Let $z = \omega_1/2$. Then $\wp'(\omega_1/2) = 0$. Similarly, we have $\wp'(\omega_2/2) = 0$ and $\wp'((\omega_1+\omega_2)/2) = 0$. Moreover, since $\wp'(z)$ is an elliptic function of order three and $\omega_1/2$, $\omega_2/2$, $(\omega_1 + \omega_2)/2$ are in the same fundamental parallelogram, they are zeros of $\wp'(z)$ with multiplicity one.

Compare with (7.8), we get

$$e_1 = \wp\left(\frac{\omega_1}{2}\right), \quad e_2 = \wp\left(\frac{\omega_2}{2}\right), \quad e_3 = \wp\left(\frac{\omega_1 + \omega_2}{2}\right).$$

Compare with (7.6), we get

$$e_1 + e_2 + e_3 = 0, \quad e_1e_2 + e_2e_3 + e_3e_2 = -\frac{g_2}{4}, \quad e_1e_2e_3 = \frac{g_3}{4}.$$

In order to factorize $\wp(z)$, we introduce two functions

$$\zeta(z) = \frac{1}{z} + \sum_{\omega\neq 0}\left(\frac{1}{z-\omega} + \frac{1}{\omega} + \frac{z}{\omega^2}\right)$$

$$\sigma(z) = z\prod_{\omega\neq 0}\left(1 - \frac{z}{\omega}\right)e^{\frac{z}{\omega}+\frac{z^2}{2\omega^2}}.$$

The convergence of $\zeta(z)$, $\sigma(z)$ is obvious. And $\zeta(z)$ is a meromorphic function, $\sigma(z)$ is an entire function. Since $\zeta(z)$ converges uniformly on every compact subset of $\mathbb{C}$, differentiating $\zeta(z)$ term by term shows that

$$\wp(z) = -\zeta'(z).$$

By the periodicity of $\wp(z)$, we have

$$\zeta'(z + \omega_1) = \zeta'(z).$$

Integrating $\zeta(z + \omega_1)$ shows that

$$\zeta(z + \omega_1) = \zeta(z) + \eta_1$$

where η_1 is a constant. Letting $z = -\omega_1/2$, we have

$$\zeta\left(\frac{\omega_1}{2}\right) = -\zeta\left(\frac{\omega_1}{2}\right) + \eta_1, \quad \eta_1 = 2\zeta\left(\frac{\omega_1}{2}\right).$$

Similarly

$$\zeta(z+\omega_2) = \zeta(z) + \eta_2$$

where $\eta_2 = 2\zeta(\omega_2/2)$. We get the following relation between η_1, η_2 and ω_1, ω_2

$$\eta_2\omega_1 - \eta_1\omega_2 = 2\pi i \tag{7.9}$$

by integrating $\zeta(z)$ along ∂P as in Fig. 9. This equation is called the *Legendre relation* and the numbers η_1, η_2 are called *basic quasi periods* of ζ.

It is easy to verify that the entire function $\sigma(z)$ satisfies

$$\frac{\sigma'(z)}{\sigma(z)} = \zeta(z)$$

and

$$\sigma(z+\omega_1) = -\sigma(z)e^{\eta_1(z+\frac{\omega_1}{2})}$$

$$\sigma(z+\omega_2) = -\sigma(z)e^{\eta_2(z+\frac{\omega_2}{2})}.$$

Now we are ready for the main theorem.

Theorem 7.6 *Let $f(z)$ be an elliptic function of order n. Assume that $a_1, \cdots, a_n$ are zeros and $b_1, \cdots, b_n$ are poles of $f(z)$ in the fundamental parallelogram (with respect to a point) each appearing as many times as its multiplicity. Then there is a complex constant c such that*

$$f(z) = c\prod_{k=1}^{n}\frac{\sigma(z-a_k)}{\sigma(z-b_k)}. \tag{7.10}$$

This theorem tells us that any elliptic function can be expressed by Weierstrass σ-function.

Proof of Theorem 7.6 Since the sum of zeros and the sum of poles of $f(z)$ differ by a period, we can always make them equal by selecting suitable representatives of these zeros and poles. Let

$$F(z) = \prod_{k=1}^{n}\frac{\sigma(z-a_k)}{\sigma(z-b_k)}.$$

Then $F(z)$ is a meromorphic function with zeros a_k and poles b_k $(k = 1, \cdots, n)$. Since

$$\sigma(z + \omega_1 - a_k) = -\sigma(z - a_k)e^{\eta_1(z - a_k + \frac{\omega_1}{2})},$$

$$\sigma(z + \omega_2 - b_k) = -\sigma(z - b_k)e^{\eta_1(z - b_k + \frac{\omega_1}{2})},$$

we have $F(z + \omega_1) = F(z)$ and similarly $F(z + \omega_2) = F(z)$. Thus $F(z)$ has the same periods as $f(z)$. Therefore $f(z) = cF(z)$ for a constant c since the quotient of $f(z)$ and $F(z)$ has no pole. The theorem is proved.

As a special case of this theorem, for any $u \in \mathbb{C}$ and $u \notin L$,

$$\wp(z) - \wp(u) = -\frac{\sigma(z + u)\sigma(z - u)}{\sigma^2(z)\sigma^2(u)}. \tag{7.11}$$

We leave the verification of this equation to the reader as an exercise.

7.3 The Jacobi Elliptic Functions

In the last section, we expressed $\wp'^2(z)$ as a product of three factors $\wp(z) - e_k, k = 1, 2, 3$. Actually, each of the factors can be written as the square of a single valued meromorphic function. Indeed, denote $\omega_3 = \omega_1 + \omega_2$ and $\eta_3 = \eta_1 + \eta_2$, then we have

$$\begin{aligned}\wp(z) - e_k &= \left(\frac{\sigma(z - \frac{\omega_k}{2})e^{\eta_k z/2}}{\sigma(z)\sigma(\frac{\omega_k}{2})}\right)^2 \\ &= \left(\frac{\sigma(z + \frac{\omega_k}{2})e^{-\eta_k z/2}}{\sigma(z)\sigma(\frac{\omega_k}{2})}\right)^2 = \left(\frac{\sigma_k(z)}{\sigma(z)}\right)^2, \quad k = 1, 2, 3, \end{aligned} \tag{7.12}$$

where

$$\sigma_k(z) = -\frac{\sigma\left(z - \frac{\omega_k}{2}\right)e^{\eta_k z/2}}{\sigma\left(\frac{\omega_k}{2}\right)} = \frac{\sigma\left(z + \frac{\omega_k}{2}\right)e^{-\eta_k z/2}}{\sigma\left(\frac{\omega_k}{2}\right)}, \quad k = 1, 2, 3. \tag{7.13}$$

Obviously, these three functions are meromorphic functions and it is easy to check that they are even. Also, by the Legendre relation, we have

$$\sigma_k(z + \omega_k) = -\sigma_k(z)e^{\eta_k(z + \omega_k/2)}, \quad k = 1, 2, 3, \tag{7.14}$$

and

$$\sigma_k(z + \omega_h) = \sigma_k(z)e^{\eta_k(z + \omega_h/2)}, \tag{7.15}$$

where $k \neq h$, $k, h = 1, 2, 3$. Moreover, the relation with $\wp(z)$ is

$$\wp'(z) = -2\frac{\sigma_1(z)\sigma_2(z)\sigma_3(z)}{\sigma^3(z)}. \tag{7.16}$$

Let $\lambda = \sqrt{e_1 - e_2}$ and $t = z/\lambda$. We define the Jacobi elliptic functions by

$$\operatorname{sn} z = \lambda \frac{\sigma(t)}{\sigma_2(t)}, \tag{7.17}$$

$$\operatorname{cn} z = \frac{\sigma_1(t)}{\sigma_2(t)}, \tag{7.18}$$

$$\operatorname{dn} z = \frac{\sigma_3(t)}{\sigma_2(t)}. \tag{7.19}$$

Since $\sigma_k(z), k = 1, 2, 3$ are even functions and $\sigma(z)$ is odd, it follows that $\operatorname{sn} z$ is an odd function, $\operatorname{cn} z$ and $\operatorname{dn} z$ are even functions. Moreover, $\operatorname{sn} 0 = 0$ and $\operatorname{cn} 0 = \operatorname{dn} 0 = 1$.

Since

$$\operatorname{sn}(z + \lambda\omega_1) = -\operatorname{sn} z, \quad \operatorname{sn}(z + \lambda\omega_2) = \operatorname{sn} z,$$

the periods of $\operatorname{sn} z$ are $2\lambda\omega_1$ and $\lambda\omega_2$. Similarly, the periods of $\operatorname{cn} z$ are $2\lambda\omega_1$ and $\lambda\omega_3$, the periods of $\operatorname{dn} z$ are $\lambda\omega_1$ and $2\lambda\omega_2$.

Let $K_1 = (1/2)\lambda\omega_1$ and $iK_2 = (1/2)\lambda\omega_2$. Then the poles of these three functions are $2mK_1 + (2n+1)iK_2$ where m, n are arbitrary integers. More precisely, the poles of $\operatorname{sn} z$ are $iK_2, 2K_1 + iK_2$, the poles of $\operatorname{cn} z$ are $2K_1 + iK_2, 4K_1 + iK_2$ and the poles of $\operatorname{dn} z$ are $iK_2, 3iK_2$. The zeros of $\operatorname{sn} z$ are $0, 2K_1$, the zeros of $\operatorname{cn} z$ are $K_1, 3K_1$, the zeros of $\operatorname{dn} z$ are $K_1 + iK_2, K_1 + 3iK_2$. Thus Jacobi elliptic functions have order two and they have two simple zeros and two simple poles on P.

Weierstrass functions $\wp(z)$, $\zeta(z)$, $\sigma(z)$, functions $\sigma_1(z)$, $\sigma_2(z)$, $\sigma_3(z)$ and Jacobi elliptic functions are all dependent on the base periods ω_1 and ω_2. By their definitions, we have

$$\wp(z)(lz, l\omega_1, l\omega_2) = \frac{1}{l^2}\wp(z, \omega_1, \omega_2),$$

$$\zeta(lz, l\omega_1, l\omega_2) = \frac{1}{l}\zeta(z, \omega_1, \omega_2),$$

$$\sigma(lz, l\omega_1, l\omega_2) = \frac{1}{l}\sigma(z, \omega_1, \omega_2),$$

$$\sigma_k(lz, l\omega_1, \omega_2) = \sigma_k(z, \omega_1, \omega_2), \quad k = 1, 2, 3$$

for any $l \in \mathbb{C}$ and $l \neq 0$. When the ratio $\tau = \omega_2/\omega_1$ is given, different choices of ω_1 and ω_2 define different $\wp(z)$, $\zeta(z)$, $\sigma(z)$, but they define the same Jacobi elliptic functions $\operatorname{sn} z$, $\operatorname{cn} z$, $\operatorname{dn} z$. This is a important difference between Weierstrass theory and Jacobi theory.

Eliminating $\wp(z)$ from equations $\wp(z) - e_k = \sigma_k^2(z)/\sigma^2(z)$, $k = 1, 2, 3$ we have

$$(e_1 - e_2)\frac{\sigma^2(z)}{\sigma_2^2(z)} + \frac{\sigma_1^2(z)}{\sigma_2^2(z)} = 1,$$

$$(e_3 - e_2)\frac{\sigma^2(z)}{\sigma_2^2(z)} + \frac{\sigma_3^2(z)}{\sigma_2^2(z)} = 1.$$

Let $k_0^2 = (e_3 - e_2)/(e_1 - e_2)$. Then we have

$$\operatorname{sn}^2 z + \operatorname{cn}^2 z = 1, \tag{7.20}$$

$$k_0^2 \operatorname{sn}^2 z + \operatorname{dn}^2 z = 1. \tag{7.21}$$

The number k_0 is called a *modular constant* and it is uniquely determined up to a sign by the ratio $\tau = \omega_2/\omega_1$.

Now we find the differential equation that $\operatorname{sn} z$ satisfies. Differentiating

$$\wp\left(\frac{z}{\lambda}\right) - e_2 = \frac{\lambda^2}{\operatorname{sn}^2 z}$$

we have

$$\frac{1}{\lambda}\wp'\left(\frac{z}{\lambda}\right) = -2\lambda^2\frac{\operatorname{sn}' z}{\operatorname{sn}^3 z}.$$

By the equation (7.16) and the definition of Jacobi elliptic functions, we have

$$\operatorname{sn}' z = \operatorname{cn} z \cdot \operatorname{dn} z. \tag{7.22}$$

Differentiating (7.20) and (7.21), we get

$$\operatorname{cn} z \operatorname{cn}' z + \operatorname{sn} z \operatorname{sn}' z = 0,$$

$$\operatorname{dn} z \operatorname{dn}' z = -k^2 \operatorname{sn} z \operatorname{sn}' z.$$

Also, by (7.22),

$$\mathrm{cn}' z = -\,\mathrm{sn}\, z \cdot \mathrm{dn}\, z,$$

$$\mathrm{dn}' z = -k^2\, \mathrm{cn}\, z \cdot \mathrm{sn}\, z.$$

Therefore, the differential equation which $\mathrm{sn}\, z$ satisfies is

$$\mathrm{sn}'^2 z = \mathrm{cn}^2\, z \cdot \mathrm{dn}^2\, z = (1 - \mathrm{sn}^2\, z)(1 - k^2\, \mathrm{sn}^2\, z). \tag{7.23}$$

Let $w = \mathrm{sn}\, z$. Then the above equation becomes

$$\frac{dw}{dz} = \sqrt{(1 - w^2)(1 - k^2 w^2)},$$

or equivalently,

$$dz = \frac{dw}{\sqrt{(1 - w^2)(1 - k^2 w^2)}}.$$

Moreover,

$$z = F(w) = \int_0^w \frac{du}{(1 - u^2)(1 - k^2 u^2)} \tag{7.24}$$

This is called an *elliptic integral* and hence $\mathrm{sn}\, z$ is an inverse function of this integral. By Schwarz-Christoff formula, (7.24) maps conformally the upper half w-plane to a rectangle on z-plane when $0 < k < 1$.

7.4 The Modular Function

Any two bases of the period module of an elliptic function (as defined in Section 7.1) can be transformed to each other by an integral matrix with determinant 1, in other words, by an element in $SL_2(\mathbb{Z})$ which we called the *modular group*.

The linear transformation between two bases (ω_1, ω_2) and (ω_1', ω_2') induces a fractional linear transformation between $\tau_1 = \omega_2/\omega_1$ and $\tau_2 = \omega_2'/\omega_1'$ in the upper half plane. We will show that any basis can be transformed into a basis whose $\tau = \omega_2/\omega_1$ lies in a fundamental domain. To achieve this goal, we study the action of $SL_2(\mathbb{Z})$ on the upper half plane.

Let z_1 and z_2 be two points on the upper half plane. If there is an element of $SL_2(\mathbb{Z})$, i.e. a modular transformation, that induces a map from z_1 to z_2, then we say that z_1 and z_2 are *congruent* (with respect to

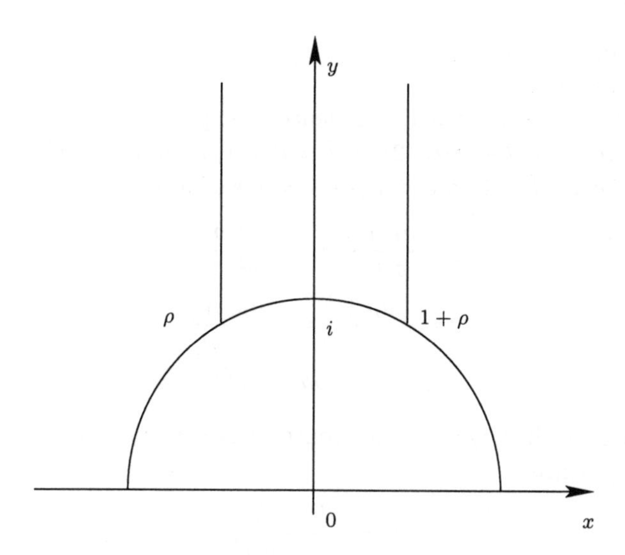

Fig. 11

the modular group), and we write $z_1 \sim z_2$. Obviously, this is an equivalent relationship, namely, it is reflexive, symmetric, and transitive.

Let D be a region on the upper half plane defined by

$$\begin{aligned} D = &\{z = x + iy| \ -\frac{1}{2} \le x < \frac{1}{2}, \\ &x^2 + y^2 > 1, \text{ if } x > 0; \ x^2 + y^2 \ge 1, \text{ if } x \le 0\}. \end{aligned}$$

Then there are no two points in D congruent to each other. Indeed, if $z_1, z_2 \in D$ (denote $z_1 = x_1 + iy_1, z_2 = x_2 + iy_2$) are distinct points and

$$z_2 = \frac{az_1 + b}{cz_1 + d},$$

where $a, b, c, d \in \mathbb{Z}$, $ad - bc = 1$, then

$$y_2 = \frac{y_1}{|cz_1 + d|^2}.$$

Since

$$|cz_1 + d|^2 = (cx_1 + d)^2 + (cy_1)^2 = c^2(x_1^2 + x_2^2) + 2cdx_1 + d^2 \ge c^2 - |cd| + d^2,$$

we see that, except the following three cases

(1) $c = \pm 1, \quad d = 0,$

(2) $c = 0, \quad d = \pm 1$,

(3) $c = d = 1$,

we always have $|cz_1 + d|^2 > 1$, and hence $y_2 < y_1$.

Denote $\rho = -1/2 + i(\sqrt{3}/2)$. If $c = d = 1$, then $|cz_1 + d|^2 = 1$ only as $z_1 = \rho$. Since $a - b = 1$ and $\rho^2 + \rho + 1 = 0$, we have

$$\begin{aligned} z_2 &= \frac{a\rho + b}{\rho + 1} = -\frac{a\rho + b}{\rho^2} \\ &= -\frac{a\rho + b}{\bar{\rho}} = -\frac{a\rho^2 + b\rho}{\rho\bar{\rho}} \\ &= -a\rho^2 - b\rho = -\rho^2 + b. \end{aligned}$$

Thus, if $z_2 \in D$, then $z_2 = \rho$. This contradicts with $z_2 \neq \rho$.

Similarly, since

$$z_1 = \frac{dz_2 - b}{-cz_2 + a},$$

we see that, except the following two cases

(1) $c = \pm 1, \quad a = 0$,

(2) $c = 0, \quad a = \pm 1$,

we always have $y_1 < y_2$. Since $y_1 < y_2$ and $y_2 < y_1$ will not occur simultaneously, we only need to study the case $c = 0, a = d = 1$ and the case $c = 1, a = d = 0$.

For the first case, $z_2 = z_1 + b, b \neq 0$. It follows that $x_2 = x_1 + b$, $|x_2 - x_2| \geq 1$ and hence z_1, z_2 are not both in D.

For the second case, $b = -1$, i.e.

$$z_2 = -\frac{1}{z_1}.$$

If $|z_1| > 1$, then $|z_2| < 1$, and $z_2 \notin D$. If $|z_2| > 1$, then $z_1 \notin D$. If $|z_1| = |z_2| = 1$, then z_1 must be on the arc from ρ to i on ∂D. Thus z_2 must be on the arc from $\rho + 1$ to i. Therefore, z_2 is in D only if $z_2 = i$. But then $z_1 = i = z_2$. This contradicts with the assumption.

Now we verify another fact. In the region

$$\{z = x + iy| \quad -\frac{1}{2} \leq x < \frac{1}{2}, \quad y > \gamma \quad (\gamma > 0)\},$$

the number of points congruent to a fixed point is finite.

Let $z_2 = (az_1 + b)/(cz_1 + d)$. Then

$$y_2 = \frac{y_1}{|cz_1 + d|^2} = \frac{y_1}{(cx_1 + d)^2 + c^2 y_1^2}.$$

If $y_2 \geq \gamma$, then the number of pairs of integers c, d that satisfy the inequality $(cx_1 + d)^2 + c^2y^2 \leq y/\gamma$ is finite. Suppose c', d' is such a pair with c'/d' reduced. Then the solution set of the equation $ad' - bc' = 1$ is $\{a = a' + mc', b = b' + md'\}$ where$\{a', b'\}$ is a fixed solution and satisfies $a'd' - b'c' = 1$, m is an arbitrary integer. Thus,

$$z_2 = \frac{az_1 + b}{c'z_1 + d'} = \frac{a'z_1 + b'}{c'z_1 + d'} + m.$$

Since there is only one m such that $-1/2 \leq x < 1/2$, for a pair c', d', there is only one pair a, b such that $-1/2 \leq x < 1/2$. Hence there are only a finite number of points in the region congruent with z.

Let $z = x_0 + iy_0, y_0 > 0$ and m be a integer such that $-1/2 \leq x_0 + m < 1/2$. Suppose $z' = z + m$. If $|z'| > 1$, then $z' \in D$. If $|z'| = 1$ and z' is on the arc from ρ to i, then $z' \in D$. If z' is on the arc from $1 + \rho$ to i, then the map $-1/z$ can bring it to the arc from ρ to i. If $|z'| < 1$, then set $z'' = -1/z'$ and $y'' = y_0/|z'|^2 > y_0$, take m' such that $z''' = z'' + m'$, $-1/2 \leq x''' < 1/2$. If $z''' \notin D$, then set $z'''' = -1/z'''$. Thus we have that points $z', z''', \cdots$ are in the region $\{-1/2 \leq x < 1/2, \quad y > y_0\}$. By the fact we just verified, the number of terms in this sequence is finite. Hence any point must be congruent with a point in D. We have proved the following theorem.

Theorem 7.7 *Any point on the upper half plane is congruent with an unique point in D.*

The region D is referred to as the *fundamental domain* of $SL_2(\mathbb{Z})$.

By the definition of Jacobi elliptic functions (7.17), (7.18) and (7.19) they are dependent on

$$e_k = \wp\left(\frac{\omega_k}{2}\right), \quad k = 1, 2, 3.$$

Hence ω_1, ω_2 is a basis of M, and $\omega_3 = \omega_1 + \omega_2$. From the differential equation of $\operatorname{sn} z$ (7.23) and the relations between the three Jacobi elliptic functions (7.20) and (7.21), we see that they are dependent on the square of the modular constant

$$k_0^2 = \frac{e_3 - e_2}{e_1 - e_2}.$$

Since $e_k, k = 1, 2, 3$ are homogeneous of order -2 on ω_1 and ω_2, k_0^2 is homogeneous of order 0 in ω_1 and ω_2. Therefore

$$k_0^2 = \lambda(\tau), \quad \tau = \frac{\omega_2}{\omega_1}$$

is a function of $\tau = \omega_2/\omega_1$.

Since e_1, e_2, e_3 are distinct, we see that $\lambda(\tau)$ is a holomorphic function on the upper half plan that does not assume the values 0 and 1.

Given a basis ω_1, ω_2 of M, any basis ω_1', ω_2' of M can be determined by an element of the modular group $SL_2(\mathbb{Z})$ as follows:

$$\omega_1' = d\omega_1 + c\omega_2,$$

$$\omega_2' = b\omega_1 + a\omega_2.$$

We continue using the notation $\omega_3' = \omega_1' + \omega_2'$. This transformation induces the following fractional linear transformation from $\tau = \omega_2/\omega_1$ to $\tau' = \omega_2'/\omega_1'$:

$$\tau' = \frac{a\tau + b}{c\tau + d}.$$

To study the relationship between $\lambda(\tau')$ and $\lambda(\tau)$, we look at the relationship between $e_k = \wp(\omega_k/2)$, $k = 1, 2, 3$ and $e_k' = \wp(\omega_k'/2)$, $k = 1, 2, 3$. Since ω_1 and ω_2 are periods of $\wp(z)$, we have

$$e_1' = \wp\left(\frac{\omega_1'}{2}\right) = \wp\left(\frac{d\omega_1 + c\omega_2}{2}\right) = \wp\left(\frac{d_1\omega_1 + c_1\omega_2}{2}\right),$$

where $d_1 = 0$ if d is even and $d_1 = 1$ if d is odd, and $c_1 = 0$ if c is even and $c_1 = 1$ if c is odd. Similarly

$$e_2' = \wp\left(\frac{\omega_2'}{2}\right) = \wp\left(\frac{b\omega_1 + a\omega_2}{2}\right) = \wp\left(\frac{b_1\omega_1 + a_1\omega_2}{2}\right),$$

where $b_1 = 0$ if b is even and $b_1 = 1$ if b is odd, and $c_1 = 0$ if c is even, and $c_1 = 1$ if c is odd. Finally

$$e_3' = \wp\left(\frac{\omega_3'}{2}\right) = \wp\left(\frac{(d+b)\omega_1 + (c+a)\omega_2}{2}\right) = \wp\left(\frac{f_1\omega_1 + h_1\omega_2}{2}\right),$$

where $f_1 = 0$ if $d + b$ is even and $f_1 = 1$ if $d + b$ is odd, and $h_1 = 0$ if $c + a$ is even and $h_1 = 1$ if $c + a$ is odd. Thus the values of e_1', e_2', e_3' depends only on the parity of a, b, c, d.

From the above discussion, it is easy to see that if

$$\begin{pmatrix} a & b \\ c & d \end{pmatrix} \equiv \begin{pmatrix} 1 & 0 \\ 0 & 1 \end{pmatrix} \pmod 2$$

Then $e_1' = e_1$, $e_2' = e_2$, $e_3' = e_3$. Therefore,

$$\lambda(\tau') = \frac{e_3' - e_2'}{e_1' - e_2'} = \frac{e_3 - e_2}{e_1 - e_2} = \lambda(\tau).$$

Similarly, if

$$\begin{pmatrix} a & b \\ c & d \end{pmatrix} \equiv \begin{pmatrix} 0 & 1 \\ 1 & 0 \end{pmatrix} \pmod 2,$$

then $e_1' = e_2$, $e_2' = e_1$, $e_3' = e_3$. Therefore,

$$\lambda(\tau') = \frac{e_3' - e_2'}{e_1' - e_2'} = \frac{e_3 - e_1}{e_2 - e_1} = 1 - \frac{e_3 - e_2}{e_1 - e_2} = 1 - \lambda(\tau).$$

It is easy to verify that any matrix in $SL_2(\mathbb{Z})$ is congruent to one of the following matrices mod 2:

$$\begin{pmatrix} 1 & 0 \\ 0 & 1 \end{pmatrix}, \begin{pmatrix} 0 & 1 \\ 1 & 0 \end{pmatrix}, \begin{pmatrix} 1 & 1 \\ 0 & 1 \end{pmatrix}, \begin{pmatrix} 1 & 1 \\ 1 & 0 \end{pmatrix}, \begin{pmatrix} 1 & 0 \\ 1 & 1 \end{pmatrix}, \begin{pmatrix} 0 & 1 \\ 1 & 1 \end{pmatrix}.$$

It is also easy to verify that

$$\lambda(\tau') = \frac{\lambda(\tau)}{\lambda(\tau) - 1},$$

if

$$\begin{pmatrix} a & b \\ c & d \end{pmatrix} \equiv \begin{pmatrix} 1 & 1 \\ 0 & 1 \end{pmatrix} \pmod 2.$$

The rest of the cases are left as an exercise for the reader.

If a meromorphic function is invariant under a subgroup of $\operatorname{Aut}\mathbb{C}^* = SL(2,\mathbb{C})/\{\pm I\}$ (see Section 3.3), then it is called an *automorphic function*. If it is automorphic with respect to a subgroup of the modular group, then it is called a *modular function*.

Since all matrices in $SL_2(\mathbb{Z})$ that satisfy

$$\begin{pmatrix} a & b \\ c & d \end{pmatrix} \equiv \begin{pmatrix} 1 & 0 \\ 0 & 1 \end{pmatrix} \pmod 2$$

form a subgroup of $SL_2(\mathbb{Z})$, $\lambda(\tau)$ is a modular function. This subgroup is called the *congruence subgroup* mod 2. It is generated by the transformations

$$\tau \longrightarrow \tau + 2, \qquad \tau \longrightarrow \frac{\tau}{1 - 2\tau}.$$

Thus we have

$$\lambda(\tau+2)=\lambda(\tau), \qquad \lambda\left(\frac{\tau}{1-2\tau}\right)=\lambda(\tau).$$

Moreover, the matrix

$$\begin{pmatrix}0 & 1\\ -1 & 0\end{pmatrix}\equiv\begin{pmatrix}0 & 1\\ 1 & 0\end{pmatrix} \pmod 2$$

induces the transformation

$$\tau \longrightarrow -\frac{1}{\tau}.$$

Hence

$$\lambda\left(-\frac{1}{\tau}\right)=1-\lambda(\tau).$$

The matrix

$$\begin{pmatrix}1 & 1\\ 0 & 1\end{pmatrix}\equiv\begin{pmatrix}1 & 1\\ 0 & 1\end{pmatrix} \pmod 2$$

induces the transformation

$$\tau \longrightarrow \tau+1.$$

Hence

$$\lambda(\tau+1)=\frac{\lambda(\tau)}{\lambda(\tau)-1}.$$

We conclude this section by mentioning that modular function have many more useful applications. It was used, for example, in the original proof of the Picard theorem (cf. Greene and Krantz [2]).

Chapter 8

The Riemann ζ-Function and The Prime Number Theorem

8.1 The Gamma Function

In this last chapter of the book, we briefly introduce the Riemann ζ-function along with some of its properties. We also prove one of the most important theorems in number theory, the prime number theorem. The reader will see the power of the Riemann ζ-function. In fact, the Riemann ζ-function not only had been used to proved the prime number theorem, but also played a crucial role in solving many major problems in number theory.

Before the Riemann ζ-function is introduced, we need to study the Gamma function. The function

$$\Gamma(z) = \int_0^\infty e^{-t} t^{z-1}\, dt, \tag{8.1}$$

where $z \in \mathbb{C}$, defined by an infinite integral, is called the *Gamma function*. This is the function $\Gamma(x)$ in calculus when $z = x > 0$. Thus, $\Gamma(z)$ is an extension of $\Gamma(x)$ from the real axis to the complex plane. Since $|e^{-t}t^{z-1}| = e^{-t}t^{x-1}$ for $z = x + iy$, the integral (8.1) converges absolutely for $\operatorname{Re} z = x > 0$. Thus, $\Gamma(z)$ is well defined and is holomorphic on $\operatorname{Re} z > 0$. Since

$$\begin{aligned}
\Gamma(z+1) &= \int_0^\infty e^{-t} t^z \, dt \\
&= -\int_0^\infty t^z \, de^{-t} \\
&= -t^z e^{-t}\Big|_0^\infty + z\int_0^\infty e^{-t}t^{z-1}\, dt \\
&= z\Gamma(z),
\end{aligned}$$

we have

$$\Gamma(z+n) = z(z+1)(z+n-1)\Gamma(z), \quad \operatorname{Re} z > 0. \tag{8.2}$$

Using (8.2), $\Gamma(z)$ can be analytically continued to the left half plane:

$$\Gamma(z) = \frac{\Gamma(z+n)}{z(z+1)\cdots(z+n-1)}, \quad \operatorname{Re} z > -n.$$

Since $\Gamma(z+n)$ is holomorphic on $\operatorname{Re} z > -n$, the right hand side of the above equation is holomorphic on the half plane $\operatorname{Re} z > -n$ except $0, -1, -2, \cdots, -(n-1)$, which are simple poles of the function. Thus $\Gamma(z)$ is analytically continued to the entire plane since n is arbitrary. The points $z = -n$, $n = 0, 1, 2, \cdots$ are simple poles of $\Gamma(z)$ and their residues are

$$\operatorname{Res}(\Gamma(z), -n) = \frac{(-1)^n}{n!}$$

Now, we give the Gauss formula and the Weierstrass formula of $\Gamma(z)$. Suppose that

$$f_n(z) = \int_0^n \left(1 - \frac{t}{n}\right)^n t^{z-1}\, dt, \quad \operatorname{Re} z > 0.$$

Let $x = t/n$. Performing integration by parts n times, we have

$$f_n(z) = n^z \int_0^1 (1-x)^n x^{z-1}\, dx = \frac{n! n^z}{z(z+1)\cdot(z+n)}.$$

On the other hand, we claim that

$$\lim_{n\to\infty} f_n(z) = \Gamma(z), \quad \operatorname{Re} z > 0.$$

That is

$$\begin{aligned}
\lim_{n\to\infty} &\left[\int_0^n e^{-t} t^{z-1}\, dt - \int_0^n \left(1 - \frac{t}{n}\right)^n t^{z-1}\, dt\right] \\
&= \lim_{n\to\infty} \int_0^n e^{-t} t^{z-1} \left[1 - e^t \left(1 - \frac{t}{n}\right)^n\right] dt = 0.
\end{aligned}$$

First, we verify that for $0 \le t \le n$, $n > 1$

$$0 < 1 - e^t \left(1 - \frac{t}{n}\right)^n < \frac{et^2}{4n}. \tag{8.3}$$

In fact, since

$$\left[e^t\left(1-\frac{t}{n}\right)^n\right]' = -\frac{1}{n}e^t t\left(1-\frac{t}{n}\right)^{n-1},$$

we have

$$1-e^t\left(1-\frac{t}{n}\right)^n = \frac{1}{n}\int_0^t e^x x\left(1-\frac{x}{n}\right)^{n-1} dx.$$

The maximal value of $e^x(1-x/n)^{n-1}$ is attained at $x=1$ and

$$e^x\left(1-\frac{x}{n}\right)^{n-1}\bigg|_{x=1} = e\left(1-\frac{1}{n}\right)^{n-1} = \frac{e}{\left(1+\dfrac{1}{n-1}\right)^{n-1}} \le \frac{e}{2}, \quad n>1.$$

So

$$1-e^t\left(1-\frac{t}{n}\right)^n \le \frac{e}{2n}\int_0^t x\,dx \le \frac{et^2}{4n}, \quad n>1.$$

Furthermore,

$$e^{-x} \ge 1-x, \quad e^{-\frac{t}{n}} \ge 1-\frac{1}{n}, \quad e^{-t} \ge \left(1-\frac{t}{n}\right)^n$$

for $x>0$. Thus,

$$1-e^t\left(1-\frac{t}{n}\right)^n > 0.$$

By (8.3), we have

$$\begin{aligned}\left|\int_0^n e^{-t}t^{z-1}\left[1-e^t\left(1-\frac{t}{n}\right)^n\right]dt\right| &\le \frac{e}{4n}\int_0^n e^{-t}t^{x+1}\,dt \\ &\le \frac{e}{4n}\Gamma(x+2).\end{aligned}$$

Therefore

$$\lim_{n\to\infty}\int_0^n e^{-t}t^{z-1}\left[1-e^t\left(1-\frac{t}{n}\right)^n\right]dt = 0$$

and the Gauss formula

$$\Gamma(z) = \lim_{n\to\infty}\frac{n!n^z}{z(z+1)\cdots(z+n)} \tag{8.4}$$

follows.

Since $\Gamma(z)$ is analytically continued to the left half plane, by the uniqueness theorem, (8.4) also holds on the left half plane, and $\Gamma(z)$ has poles at $z = 0, -1, -2, \cdots$.

Since

$$\frac{n!n^z}{z(z+1)\cdots(z+n)} = \frac{e^{z\log n}}{z(1+z)\left(1+\frac{z}{2}\right)\cdots\left(z+\frac{z}{n}\right)}$$
$$= \frac{e^{z\log z}}{z\prod_{k=1}^{n}\left(1+\frac{z}{k}\right)}$$
$$= \frac{e^{z(\log n - \sum_{k=1}^{n}\frac{1}{k})}}{z\prod_{k=1}^{n}\left(1+\frac{z}{k}\right)e^{-\frac{z}{k}}}.$$

We also have

$$\frac{1}{\Gamma(z)} = \lim_{n\to\infty}\frac{z(z+1)\cdots(z+n)}{n!n^z} = e^{\gamma z}z\prod_{n=1}^{\infty}\left(1+\frac{z}{n}\right)e^{-\frac{z}{n}}, \tag{8.5}$$

where γ is the Euler constant.

It is easy to verify that

$$e^{\gamma z}z\prod_{n=1}^{\infty}\left(1+\frac{z}{n}\right)e^{-\frac{z}{n}}$$

is an entire function and has zeros $z = 0, -1, -2, \cdots$. The equation (8.5) is called the *Weierstrass formula*.

It is obvious that $\Gamma(z) \neq 0$. It follows from (8.5) that

$$\frac{1}{\Gamma(-z)} = -e^{-\gamma z}z\prod_{n=1}^{\infty}\left(1-\frac{z}{n}\right)e^{\frac{z}{n}}.$$

By Exercise 3, problem 6(v) in Chapter 3, we have

$$\frac{1}{\Gamma(z)\Gamma(-z)} = -z^2\prod_{n=1}^{\infty}\left(1-\frac{z^2}{n^2}\right) = \frac{-z}{\pi}\sin\pi z.$$

Also $\Gamma(1-z) = -z\Gamma(-z)$ by (8.2). Thus

$$\Gamma(z)\Gamma(1-z) = \frac{\pi}{\sin\pi z}. \tag{8.6}$$

The equation (8.6) is called the *Euler reflection formula*. Now we are ready to define the Riemann ζ-function.

8.2 The Riemann ζ-function

Let $s = \sigma + it$ be a complex number. The series

$$\zeta(s) = \sum_{n=1}^{\infty} \frac{1}{n^s}, \quad \sigma > 0 \tag{8.7}$$

is called the *Riemann ζ-function.*

For any $a > 1$, if $\sigma \geq a$, we have

$$\left| \sum_{n=N}^{\infty} \frac{1}{n^s} \right| \leq \sum_{n=N}^{\infty} \frac{1}{n^\sigma} \leq \sum_{n=N}^{\infty} \frac{1}{n^a}$$

where $N \in \mathbb{N}$. Since $\sum_{n=1}^{\infty} 1/n^a$ converges for $a > 1$, $\zeta(s)$ converges uniformly for $\sigma \geq a > 1$. Thus $\zeta(s)$ is holomorphic on the half plane $\sigma > 1$ since a is an arbitrary positive number that is great than 1.

How is $\zeta(s)$ connected to prime numbers?

Let $p_1, p_2, \cdots, p_n, \cdots$ be the sequence of all prime numbers in the increasing order, i.e., $p_1 = 2$, $p_2 = 3$, $p_3 = 5, \cdots$. The following infinite product theorem expresses the Riemann ζ-function by prime numbers.

Theorem 8.1 *Let $\{p_n\}$ be the increasing sequence of all prime numbers. Then*

$$\zeta(s) = \prod_{n=1}^{\infty} \left(1 - p_n^{-s}\right)^{-1}, \quad \operatorname{Re} s > 1. \tag{8.8}$$

Proof It is well known that

$$\sum_{n=1}^{\infty} |p_n^{-s}| = \sum_{n=1}^{\infty} p_n^{-\sigma}$$

converges absolutely for $\operatorname{Re} s > 1$. The Taylor series of $(1-x)^{-1}$ is also absolutely convergent for $|x| < 1$. So by Section 3.4, the infinite product on the right hand side of (8.8) converges absolutely for $\sigma > 1$. Also, for $\sigma > 1$, we have

$$\begin{aligned} \zeta(s)\left(1 - \frac{1}{2^s}\right) &= \sum_{n=1}^{\infty} \frac{1}{n^s} - \sum_{n=1}^{\infty} \frac{1}{(2n)^s} \\ &= 1 + \frac{1}{3^s} + \frac{1}{5^s} + \cdots . \end{aligned}$$

and

$$\zeta(s)\left(1-\frac{1}{2^s}\right)\left(1-\frac{1}{3^s}\right)=1+\frac{1}{5^s}+\frac{1}{7^2}+\cdots.$$

The series on the right hand side does not have terms that contain 2 and 3 as their factors. In general,

$$\zeta(s)\left(1-\frac{1}{2^s}\right)\left(1-\frac{1}{3^s}\right)\cdots\left(1-\frac{1}{p_n^s}\right)=1+\frac{1}{m_1^s}+\frac{1}{m_2^s}\cdots$$

where the prime factors of $m_1, m_2, \cdots$ are greater than p_n. Thus

$$\left|\zeta(s)\left(1-\frac{1}{2^s}\right)\cdots\left(1-\frac{1}{p_n^s}\right)\right|\leq 1+\frac{1}{(p_n+1)^\sigma}+\frac{1}{(p_n+2)^\sigma}+\cdots.$$

Obviously, $p_n \geq n$ and the remainder of $\sum_{n=1}^{\infty}\frac{1}{n^\sigma}$ tends to 0 as $n\to\infty$. Therefore

$$\lim_{n\to\infty}\zeta(s)\prod_{n=1}^{n}\left(1-\frac{1}{p_n^s}\right)=1$$

and (8.8) follows. Equation (8.8) is called the *Euler formula*. It is easy to see that $\zeta(s)\neq 0$, $\sigma>1$.

Now we study the analytical continuation of the Riemann ζ-function. From (8.1) we have

$$\Gamma(s)=\int_0^\infty e^{-x}x^{s-1}\,dx$$

for $\operatorname{Re}s>0$. Substitute nx for x, we get that

$$\frac{\Gamma(s)}{n^s}=\int_0^\infty e^{-nx}x^{s-1}\,dx.$$

Take the sum on both side of this equation with respect to n for $\operatorname{Re}s>1$, we have

$$\zeta(s)\Gamma(s)=\sum_{n=1}^{\infty}\int_0^\infty e^{-nx}x^{s-1}\,dx.$$

Since

$$\int_0^\infty\sum_{n=1}^{\infty}|e^{-nx}x^{s-1}|\,dx=\int_0^\infty\sum_{n=1}^{\infty}e^{-nx}x^{\sigma-1}\,dx=\int_0^\infty\frac{x^{\sigma-1}}{e^x-1}\,dx$$

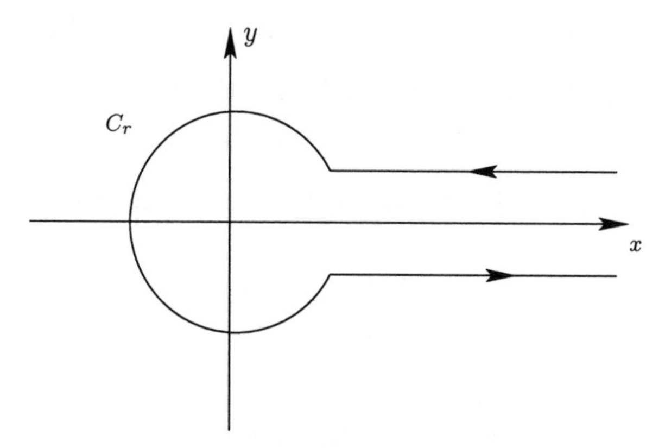

Fig. 12

converges, we have

$$\zeta(s)\Gamma(s) = \int_0^\infty \sum_{n=1}^\infty e^{-nx} x^{s-1}\, dx$$

$$= \int_0^\infty \frac{x^{s-1}}{e^x - 1}\, dx, \quad \operatorname{Re} s > 1. \tag{8.9}$$

Consider the integral

$$F(s) = \int_C \frac{z^{s-1}}{e^z - 1}\, dz \tag{8.10}$$

where C consists of arc

$$C_r : re^{i\theta}, \quad \frac{\pi}{6} \le \theta \le \frac{5\pi}{6},$$

and two half lines:

$$y = \pm\frac{1}{2}r, \quad \frac{\sqrt{3}}{2}r \le x < \infty.$$

The curve C is oriented as in Fig. 12. It is easy to see that the integral does not depend on r if $r < 2\pi$. Thus,

$$F(s) = \int_{C_r} \frac{z^{s-1}}{e^z - 1}\, dz - \int_{\frac{\sqrt{3}}{2}r}^\infty \frac{(x + \frac{ir}{2})^{s-1}}{e^{x+\frac{ir}{2}} - 1}\, dx + \int_{\frac{\sqrt{3}}{2}}^\infty \frac{(x - \frac{ir}{2})^{s-1}}{e^{x-\frac{ir}{2}} - 1}\, dx.$$

Obviously the first integral on the right hand side is an entire function. The second integral is also an entire function since it converges uniformly on any disc $|z| < R$. Thus, $F(s)$ is an entire function. For $z \in C_r$,

$$\begin{aligned}|z^{s-1}| &= |e^{(s-1)\log z}| = e^{\mathrm{Re}\{(\sigma-1+it)(\log r + i\arg z)\}}\\ &= e^{(\sigma-1)\log r - t\arg z} \leq r^{\sigma-1}e^{2\pi t+1}\end{aligned}$$

and $|e^z - 1| \geq A(z)$, where A is a constant. So for any fixed s ($\sigma > 1$), we have

$$\left|\frac{z^{s-1}}{e^z - 1}\right| = O(r^{\sigma-2}).$$

Thus

$$\left|\int_{C_r} \frac{z^{s-1}}{e^z - 1}\, dz\right| = O(r^{\sigma-1})$$

tends to 0 as $r \to 0$. Hence

$$F(s) = (e^{2\pi is} - 1)\int_0^\infty \frac{x^{s-1}}{e^x - 1}\, dx.$$

By (8.9) we have

$$F(s) = (e^{2\pi is} - 1)\Gamma(s)\zeta(s)$$

for $\sigma > 1$. That is

$$\zeta(s) = \frac{1}{(e^{2\pi is} - 1)\Gamma(s)}\int_C \frac{z^{s-1}}{e^z - 1}\, dz.$$

By (8.6),

$$\begin{aligned}(e^{2\pi is} - 1)\Gamma(s) &= e^{\pi is}(e^{\pi is} - e^{-\pi is})\Gamma(s)\\ &= 2ie^{\pi is}\sin\pi s\Gamma(s) = \frac{2\pi ie^{\pi is}}{\Gamma(1-s)}\end{aligned}$$

Thus

$$\zeta(s) = \frac{e^{-\pi is}\Gamma(1-s)}{2\pi i}\int_C \frac{z^{s-1}}{e^z - 1}\, dz. \tag{8.11}$$

Since the integral on the right hand side is an entire function, the right hand side has only one pole at $s = 1$. The residue is

$$\begin{aligned}\mathrm{Res}(\zeta(s), 1) &= \frac{1}{2\pi i}\frac{F(1)}{\Gamma(1)}\\ &= \frac{1}{2\pi i}F(1) = \frac{1}{2\pi i}\int_{|z|=1}\frac{dz}{e^z - 1} = 1.\end{aligned}$$

Therefore, (8.11) is an analytic continuation of $\zeta(s)$ on the whole plane. This proves the following theorem.

Theorem 8.2 *The Riemann ζ-function can be analytically continued to a meromorphic function (8.11) on* $\mathbb{C}$. *The point* $s = 1$ *is its only pole. It is a simple pole with residue* 1.

By (8.11) we know that if n is a non-negative integer, then

$$\zeta(-n) = \frac{(-1)^n n!}{2\pi i} \int_C \frac{z^{-n-1}}{e^z - 1}\, dz = \frac{(-1)^n n!}{2\pi i} \int_{|z|=\frac{1}{2}} \frac{z^{-n-1}}{e^z - 1}\, dz. \tag{8.12}$$

Let $\varphi(z) = z/(e^z - 1)$. Then $\varphi(z) + z/2$ is an even function. Suppose its Taylor series is

$$\varphi(z) + \frac{z}{2} = \sum_{n=0} \frac{a_{2n}}{(2n)!} z^{2n}, \quad |z| < 2\pi.$$

Obviously, $a_0 = 1$. Denote $B_n = (-1)^{n-1} a_{2n}$ and call it *Bernulli number*. Then

$$\frac{1}{e^z - 1} = \frac{1}{z} - \frac{1}{2} + \sum_{n=1}^{\infty} \frac{(-1)^{n-1} B_n}{(2n)!} z^{2n-1}, \quad |z| < 2\pi. \tag{8.13}$$

By the definition of B_n, we have

$$B_1 = \frac{1}{6}, \quad B_2 = \frac{1}{30}, \quad B_3 = \frac{1}{42}, \cdots.$$

Substitute (8.13) for (8.12) we get

$$\zeta(0) = -\frac{1}{2}, \quad \zeta(-2m) = 0,$$

$$\zeta(-2m+1) = (-1)^{2m-1}(2m-1)! \frac{(-1)^{m-1} B_n}{(2m)!} = \frac{(-1)^m B_n}{2m}$$

where $m = 1, 2, \cdots$. The points $z = -2m$, $m = 1, 2, \cdots$ are called the *trivial zeros* of ζ.

Next, we derive a function equation of $\zeta(s)$. Consider

$$\psi(z) = \frac{z^{s-1}}{e^z - 1}$$

and its integral along the path c_n shown in Fig. 13. The residues of $\psi(z)$ at the poles inside c_n are

$$\mathrm{Res}(\psi(z), 2\pi i k) = (2\pi i k)^{s-1} = -i(2\pi k)^{s-1} e^{\frac{\pi i s}{2}},$$

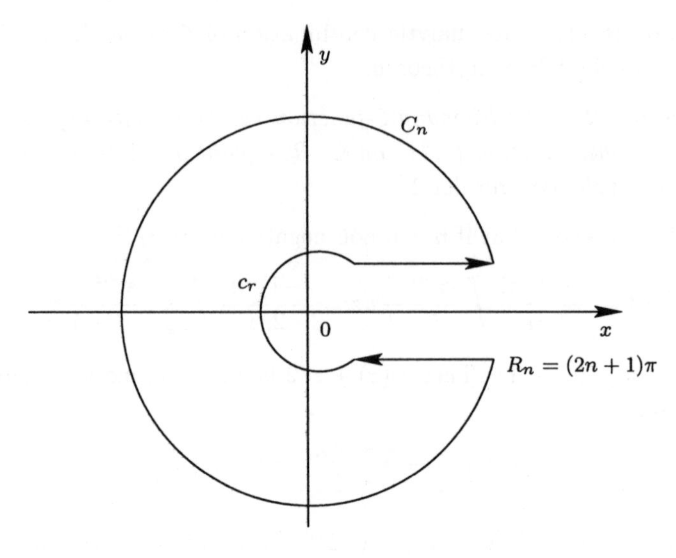

Fig. 13

$$\operatorname{Res}(\psi(z), -2\pi ik) = (-2\pi ik)^{s-1} = i(2\pi k)^{s-1}e^{\frac{3\pi i}{2}s}, \quad k = 1, 2, \cdots, n.$$

So the sum of the residues inside c_n are

$$\begin{aligned} i(e^{\frac{3\pi is}{2}} - e^{\frac{\pi is}{2}})\sum_{k=1}^{n}(2\pi k)^{s-1} &= ie^{\pi is} \cdot 2i \sin\frac{\pi s}{2}\sum_{k=1}^{n}(2\pi k)^{s-1} \\ &= -2e^{\pi is}\sin\frac{\pi s}{2}\sum_{k=1}^{n}(2\pi k)^{s-1}. \end{aligned}$$

Since $|z^{s-1}| \le R_n^{\sigma-1}e^{2\pi t}$ on $|z| = R_n$ and $1/(e^z - 1)$ is bounded,

$$\left|\int_{|z|=R_n}\psi(z)\,dz\right| = O(R_n^{\sigma}).$$

Since

$$\int_{|z|=R_n}\psi(z)\,dz$$

tends to 0 as $\sigma < 0$ and $n \to \infty$, by the residue theorem

$$\int_{|z|=R_n}\psi(z)\,dz - \int_{c_n'}\psi(z)\,dz = -4\pi ie^{\pi is}\sin\frac{\pi s}{2}\sum_{k=1}^{n}(2\pi k)^{s-1}$$

for $\sigma < 0$, where $c_n' = c_n \setminus \{|z| = r_n\}$. Let $n \to \infty$. Then

$$-\int_{c_n'} \psi(z)\,dz \to -F(z)$$

where $F(s)$ is defined by (8.10). Thus

$$F(s) = 4\pi i e^{\pi i s} \sin\frac{\pi s}{2}(2\pi)^{s-1}\zeta(1-s), \quad \sigma < 0.$$

This equation holds for all s by the fact that $F(s)$ is an entire function. It is known that

$$F(s) = (e^{2\pi i s} - 1)\zeta(s)\Gamma(s).$$

Hence

$$(e^{2\pi i s} - 1)\Gamma(s)\zeta(s) = 4\pi i e^{\pi i s} \sin\frac{\pi s}{2}(2\pi)^{s-1}\zeta(1-s).$$

That is

$$\zeta(1-s) = 2^{1-s}\pi^{-s} \cos\frac{\pi s}{2}\Gamma(s)\zeta(s), \tag{8.14}$$

or equivalently

$$\zeta(s) = 2^s \pi^{s-1} \sin\frac{\pi s}{2}\zeta(1-s)\Gamma(1-s). \tag{8.15}$$

This is the function equation of $\zeta(s)$.

Now we study the zeros of $\zeta(s)$. We already know that $\zeta(s)$ has no zeros when $\sigma > 1$. For $\sigma < 0$, by equation (8.15), Since $\Gamma(1-s)$ and $\zeta(1-s)$ does not vanish, and $\sin(\pi s/2) = 0$ only at $s = -2, -4, \cdots$, we have $\zeta(-2m) = 0$, $m = 1, 2, \cdots$ are the only zeros. Other zeros of $\zeta(s)$ are in $0 \le \sigma \le 1$.

We now show that $\zeta(s) \neq 0$ on $\sigma = 1$ and $\sigma = 0$. For the case of $\sigma = 1$, consider the function

$$\eta_\varepsilon(t) = |\zeta(1+\varepsilon)|^3|\zeta(1+\varepsilon+it)|^4|\zeta(1+\varepsilon+2it)|, \quad \varepsilon > 0 \tag{8.16}$$

Let

$$\eta_\varepsilon(t) = \prod_{n=1}^{\infty} b_n$$

where

$$b_n = |1 - p_n^{-1-\varepsilon}|^{-3}|1 - p_n^{-1-\varepsilon-it}|^{-4}|1 - p_n^{-1-\varepsilon-2it}|^{-1}.$$

From Theorem 8.1, since $\log|z| = \operatorname{Re}\log z$, we have

$$\log b_n = -3\log(1-p_n^{-1-\varepsilon}) \\ - \operatorname{Re}(4\log(1-p_n^{-1-\varepsilon-it}) + \log(1-p_n^{-1-\varepsilon-2it})).$$

Expand the above equation using the Taylor series of $\log(1-z)$ we get

$$\log b_n = \sum_{m=1}^{\infty} \frac{1}{m} p_n^{-(1+\varepsilon)m}(3 + 4\cos(mt\log p_n) + (\cos(2mt\log p_n)).$$

Since $3 + 4\cos\theta + \cos 2\theta = 2(1-\cos\theta)^2 \geq 0$, we have $\log b_n \geq 0$ i.e. $b_n \geq 1$. It follows that

$$\eta_\varepsilon(t) \geq 1, \quad \varepsilon > 0 \tag{8.17}$$

for $-\infty < t < \infty$. If there exists a $t \neq 0$ such that $\zeta(1+it) = 0$, then by Theorem 8.2 we have

$$\lim_{\varepsilon \to 0} \varepsilon\zeta(1+\varepsilon) = 1,$$

$$\lim_{\varepsilon \to 0} \frac{1}{\varepsilon}\zeta(1+\varepsilon+it) = \zeta'(1+it).$$

Substitute into (8.16), we get

$$\lim_{\varepsilon \to 0} \eta_\varepsilon(t) = 0.$$

This contradicts with (8.17). So we proved that $\zeta(1+it) \neq 0$ as $\sigma = 1$.

By (8.15) and Theorem 8.2 we know that $\zeta(s) \neq 0$ as $\sigma = 0$. So all non-trivial zeros of $\zeta(s)$ are in the belt region $0 < \operatorname{Re} s < 1$. The Riemann conjecture is: all non-trivial zeros of $\zeta(s) = \zeta(\sigma+it)$ are on the line $\sigma = 1/2$. This conjuncture is still open.

8.3 The Prime Number Theorem

Since any integer can be represented as a product of prime numbers, a major object of concern in the number theory is prime numbers. From the times of Euclid, it is well known that there are infinitely many prime numbers. Naturally, the distribution of prime numbers is one of the interesting problems in number theory. Starting from listing prime numbers by their values and getting some empirical results and conjectures, after efforts of many mathematicians over a long time, the following prime number theorem was proved

Theorem 8.3 *(the Prime Number Theorem) Let $\pi(x)$ be the number of primes not exceeding x, where $x \in \mathbb{R}$ and $x > 0$. Then*

$$\pi(x) \sim \frac{x}{\log x}, \tag{8.18}$$

as $x \to \infty$. That is

$$\lim_{x\to\infty} \frac{\pi(x)\log x}{x} = 1.$$

The proof we give here depends on the Riemann ζ-function. The "elementary proof" (a proof without using the Riemann ζ-function) was one of the difficult long standing problems in number theory. It was found at 1949 after a long search.

We introduce several number-theoretic functions first. Suppose that $x > 0$ and n is a positive integer. The *von Magoldt function* $\Lambda(n)$ is defined by

$$\Lambda(n) = \begin{cases} \log p, & \text{if } n \text{ is a power of a prime number } p, \\ 0, & \text{otherwise.} \end{cases}$$

The *Chebyshev functions* $\theta(x)$, $\psi(x)$ are defined by

$$\theta(x) = \sum_{p_n \le x} \log p_n,$$

$$\psi(x) = \sum_{n \le x} \Lambda(n) = \sum_{p^m \le x} \log p,$$

where $\{p_n\}$ is the increasing sequence of all prime numbers. From these definitions, we get that

$$\psi(x) = \theta(x) + \theta(x^{\frac{1}{2}}) + \theta(x^{\frac{1}{3}}) + \cdots$$

and

$$\psi(x) = \sum_{n\le x} \Lambda(n) = \sum_{\substack{l,m \\ n=p_m^l \le x}} \Lambda(n)$$

$$= \sum_{\substack{l=1 \\ p_m \le x}}^{[\frac{\log x}{\log p_m}]} \log p_m = \sum_{p_m \le x} \left[\frac{\log x}{\log p_m}\right] \log p_m. \tag{8.19}$$

Here $[w]$ is the integral part of w. Also by these definitions, $\pi(x)$, $\theta(x)$ and $\psi(x)$ are non-negative increasing functions. Furthermore,

$$0 \le \theta(x) \le \psi(x) \le \sum_{p_n \le x} \frac{\log x}{\log p_n} \log p_n = \pi(x) \log x, \quad x > 1$$

and $\pi(x) \le x$ as $x > 0$.

For $0 < \alpha < 1$ and $x > 1$, we have

$$\begin{aligned}\theta(x) &\ge \sum_{x^\alpha < p_n \le x} \log p_n = \sum_{x^\alpha < p_n \le x} \frac{\log p_n}{\log x^\alpha} \cdot \log x^\alpha \\ &\ge \alpha \sum_{x^\alpha < p_n \le x} \log x = \alpha(\pi(x) - \pi(x^\alpha)) \log x \\ &\ge \alpha(\pi(x) - x^\alpha) \log x.\end{aligned}$$

Thus

$$\frac{\alpha\pi(x)\log x}{x} - \frac{\alpha x^\alpha \log x}{x} \le \frac{\theta(x)}{x} \le \frac{\psi(x)}{x} \le \frac{\pi(x)}{x} \le \frac{\pi(x)\log x}{x}. \tag{8.20}$$

Letting $x \to \infty$ in (8.20) and taking upper limits, we have

$$\alpha \varlimsup_{x\to+\infty} \frac{\pi(x)\log x}{x} \le \varlimsup_{x\to+\infty} \frac{\theta(x)}{x} \le \varlimsup_{x\to+\infty} \frac{\psi(x)}{x} \le \varlimsup_{x\to+\infty} \frac{\pi(x)\log x}{x}.$$

Since α can be any number in $(0,1)$, we have

$$\varlimsup_{x\to+\infty} \frac{\pi(x)\log x}{x} = \varlimsup_{x\to+\infty} \frac{\theta(x)}{x} = \varlimsup_{x\to+\infty} \frac{\psi(x)}{x} \tag{8.21}$$

when $\alpha \to 1-0$. Taking lower limits in (8.20) and letting $\alpha \to 1-0$ we get

$$\varliminf_{x\to+\infty} \frac{\pi(x)\log x}{x} = \varliminf_{x\to\infty} \frac{\theta(x)}{x} = \varliminf_{x\to+\infty} \frac{\psi(x)}{x}. \tag{8.22}$$

From (8.21) and (8.22), we see that if the equation

$$\lim_{x\to\infty} \frac{\psi(x)}{x} = 1 \tag{8.23}$$

holds, then the Prime Number Theorem (8.18) follows.

In order to prove (8.23), consider the integral

$$\begin{aligned} f(s) &= \int_0^\infty \frac{\psi(e^t)}{e^{st}}\,dt = \int_1^\infty \frac{\psi(u)}{u^{1+s}}\,du \\ &= \sum_{n=1}^\infty \int_n^{n+1} \frac{\psi(u)}{u^{1+s}}\,du = \sum_{n=1}^\infty \sum_{m\le n} \Lambda(m) \int_n^{n+1} \frac{1}{u^{1+s}}\,ds \\ &= \frac{1}{s}\sum_{n=1}^\infty \sum_{m\le n} \Lambda(m) \left(\frac{1}{n^s} - \frac{1}{(n+1)^s}\right) \\ &= \frac{1}{s}\sum_{m=1}^\infty \Lambda(m) \sum_{n=m}^\infty \left(\frac{1}{n^s} - \frac{1}{(n+1)^s}\right) = \frac{1}{s}\sum_{n=1}^\infty \frac{\Lambda(n)}{n^s} \end{aligned} \tag{8.24}$$

where $\operatorname{Re} s > 1$. Taking the logarithmic derivative of equation (8.8) in Theorem 8.1, we have

$$\begin{aligned} \frac{\zeta'(s)}{\zeta(s)} &= \sum_{n=1}^\infty \left(\log \frac{1}{1-\frac{1}{p_n^s}}\right)' = -\sum_{n=1}^\infty \frac{\log p_n}{p_n^s} \frac{1}{1-\frac{1}{p_n^s}} \\ &= -\sum_{n=1}^\infty \log p_n \sum_{m=1}^\infty \frac{1}{p_n^{ms}} = -\sum_{n=1}^\infty \sum_{m=1}^\infty \frac{\Lambda(p_n^m)}{p_n^{ms}} = -\sum_{n=1}^\infty \frac{\Lambda(n)}{n^s} \end{aligned} \tag{8.25}$$

The last equality holds because $p_n^m, \quad m, n = 1, 2, \cdots$ are distinct positive integers and $\Lambda(k) = 0$ when the k is not of the form p_n^m. From (8.24) and (8.25) we have

$$f(s) - \frac{1}{s-1} = -\frac{1}{s}\frac{\zeta'(s)}{\zeta(s)} - \frac{1}{s-1} = -\frac{1}{s}\left(\frac{\zeta'(s)}{\zeta(s)} + \frac{1}{s-1}\right) - \frac{1}{s} \tag{8.26}$$

as $\operatorname{Re} s > 1$.

Now we consider

$$q(s) = \frac{\zeta'(s)}{\zeta(s)} + \frac{1}{s-1}.$$

Since $\zeta(s)$ is a meromorphic function, so are $1/\zeta(s)$ and $\zeta'(s)$. Hence $q(s)$ is a meromorphic function. By Theorem 8.2,

$$\zeta(s) = \frac{1}{s-1} + l(s)$$

where $l(s)$ is an entire function. We know that $\zeta(s) \neq 0$ when $\operatorname{Re} s \geq 1$, so $1 + (s-1)l(s) \neq 0$ when $\operatorname{Re} s \geq 1$. Thus

$$\frac{\zeta'(s)}{\zeta(s)} + \frac{1}{s-1} = \frac{(s-1)l'(s) + l(s)}{1 + (s-1)l(s)}.$$

That is, $q(s)$ is holomorphic on $\operatorname{Re} s \geq 1$. By (8.26), the function defined by

$$g(t) = \lim_{\sigma \to 1} \left(f(s) - \frac{1}{s-1} \right), \quad s = \sigma + it \tag{8.27}$$

converges uniformly on any compact subset of $-\infty < t < \infty$ and $g'(t)$ is continuous on $-\infty < t < \infty$. In summary, we have

Lemma 8.1 *The integral defined by (8.24)*

$$f(s) = \int_0^{\infty} \frac{\psi(e^t)}{e^{st}} \, dt$$

converges when $\operatorname{Re} s > 1$ *and*

$$g(t) = \lim_{\sigma \to 1} \left(f(s) - \frac{1}{s-1} \right)$$

converges uniformly on any compact subset of $-\infty < t < \infty$, *and* $g'(t)$ *is continues on* $-\infty < t < \infty$.

8.4 The Proof of The Prime Number Theorem

Having Lemma 8.1 as preparation, we only need to prove the important Ikehara theorem, a Tauberian theorem, in order to prove the prime number theorem (8.23).

Suppose $-\infty < x < \infty$, $\lambda > 0$. Let

$$K_\lambda(x) = \begin{cases} 1 - \dfrac{|x|}{2\lambda}, & \text{if } |x| \leq 2\lambda; \\ 0, & \text{if } |x| > 2\lambda. \end{cases}$$

$$k_\lambda(x) = \begin{cases} \dfrac{2\lambda}{\sqrt{2\pi}} \left(\dfrac{\sin \lambda x}{\lambda x} \right)^2, & \text{if } x \neq 0; \\ \dfrac{2\lambda}{\sqrt{2\pi}}, & \text{if } x = 0. \end{cases}$$

Lemma 8.2

$$\frac{1}{\sqrt{2\pi}}\int_{-\infty}^{\infty} K_\lambda(t)e^{ixt}\,dt = k_\lambda(x), \tag{8.28}$$

$$K_\lambda(x) = \frac{1}{\sqrt{2\pi}}\int_{-\infty}^{\infty} k_\lambda(t)e^{ixt}\,dt. \tag{8.29}$$

We need the following two results before we prove this lemma.

(1) If $f(x)$ has continuous first order derivative, then

$$\int_a^b f(x)e^{ixt}\,dx = O\left(\frac{1}{t}\right). \tag{8.30}$$

This can be easily verified using integration by parts.

(2)If $a < 0 < b$ and $f(x)$ has continuous second order derivative, then

$$\lim_{\omega\to\infty}\frac{1}{\pi}\int_a^b f(x)\frac{\sin\omega x}{x}\,dx = f(0). \tag{8.31}$$

Consider

$$\int_a^b (f(x)-f(0))\frac{\sin\omega x}{x}\,dx.$$

The function $(f(x)-f(0))/x$ has continuous first order derivative at 0. By (8.30) we have

$$\lim_{\omega\to\infty}\frac{1}{\pi}\int_a^b (f(x)-f(0))\frac{\sin\omega x}{x}\,dx = 0.$$

Thus

$$\begin{aligned}\lim_{\omega\to\infty}\frac{1}{\pi}\int_a^b f(x)\frac{\sin\omega x}{x}\,dx &= f(0)\lim_{\omega\to\infty}\frac{1}{\pi}\int_a^b \frac{\sin\omega x}{x}\,dx\\ &= f(0)\frac{1}{\pi}\lim_{\omega\to\infty}\int_{a\omega}^{b\omega}\frac{\sin x}{x}\,dx\\ &= f(0)\frac{1}{\pi}\int_{-\infty}^{\infty}\frac{\sin x}{x}\,dx = f(0).\end{aligned}$$

The last step uses the fact

$$\frac{1}{\pi}\int_{-\infty}^{\infty}\frac{\sin x}{x}\,dx = 1$$

from calculus.

Proof of Lemma 8.2 It is easy to see that

$$\frac{1}{\sqrt{2\pi}}\int_{-\infty}^{\infty} K_\lambda(t)e^{ixt}\,dt = \frac{2}{\sqrt{2\pi}}\int_0^{2\lambda}\left(1-\frac{t}{2\lambda}\right)\cos xt\,dt. \tag{8.32}$$

If $x = 0$, then the right hand side of the above equation is equal to $(1/\sqrt{2\pi})2\lambda$. If $x \neq 0$, then perform integration by part twice to the right hand side we get that

$$\frac{2}{\sqrt{2\pi}}\int_0^{2\lambda}\left(1-\frac{t}{2\lambda}\right)\cos xt\,dt = \frac{2\lambda}{\sqrt{2\pi}}\left(\frac{\sin\lambda x}{\lambda x}\right)^2.$$

And (8.28) follows. For (8.29), consider the integral

$$I(\omega) = \frac{1}{\sqrt{2\pi}}\int_{-\infty}^{\infty} k_\lambda(t)e^{ixt}\,dt = \frac{2}{\sqrt{2\pi}}\int_0^{\infty} k_\lambda(t)\cos xt\,dt.$$

By (8.32)

$$\begin{aligned} I(\omega) &= \frac{2}{\pi}\int_0^{\infty}\int_0^{2\lambda}\left(1-\frac{u}{2\lambda}\right)\cos ut\cos xt\,du\,dt \\ &= \frac{1}{\pi}\int_0^{2\lambda}\left(1-\frac{u}{2\lambda}\right)du\int_0^{\infty}(\cos(u+x)t+\cos(u-x)t)\,dt \\ &= \frac{1}{\pi}\int_0^{2\lambda}\left(1-\frac{u}{2\lambda}\right)\left(\frac{\sin(u+\lambda)\omega}{u+x}+\frac{\sin(u-x)\omega}{u-x}\right)du. \end{aligned} \tag{8.33}$$

If $x > 2\lambda$, then by (8.30), $\lim_{\omega\to\infty} I(\omega) = 0$. If $0 < x < 2\lambda$, then by (8.30), the limit of the first term is 0, and by (8.31), the limit of the second term is $1 - x/(2\lambda)$. Hence (8.29) follows.

Definition 8.1 If $f(x)$ is defined on $-\infty < x < \infty$ and satisfies

$$\lim_{\substack{y-x\to 0\\ x\to\infty}}(f(y)-f(x)) \geq 0, \quad (y > x), \tag{8.34}$$

then $f(x)$ is called a *slowly decreasing function*, or equivalently, for any given $\varepsilon > 0$, there exist $x_0 > 0$ and $\delta > 0$, such that

$$f(y) - f(x) > -\frac{\varepsilon}{2}$$

when $x \geq x_0$ and $0 \leq y - x < 2\delta$.

Lemma 8.3 *Let $f(x)$ be a slowly decreasing function and $|f(x)| < M$, $(-\infty < x < \infty)$. If*

$$\lim_{x\to\infty} \frac{1}{\sqrt{2\pi}} \int_{-\infty}^{\infty} k_\lambda(x-t)f(t)\,dt = L$$

for all $\lambda > 0$, then $f(x) \to L$.

Proof It is well known in calculus that

$$\int_0^\infty \frac{\sin x}{x}\,dx = \frac{\pi}{2}.$$

Thus

$$\begin{aligned}
\frac{1}{\sqrt{2\pi}} \int_{-\infty}^{\infty} k_\lambda(x)\,dx &= \frac{1}{\pi} \int_{-\infty}^{\infty} \left(\frac{\sin x}{x}\right)^2 dx \\
&= \frac{2}{\pi} \int_0^\infty \sin^2 x\, d(-\frac{1}{x}) \\
&= \frac{2}{\pi} \int_0^\infty \frac{\sin x}{x}\,dx = 1.
\end{aligned}$$

Without loss of generosities, we may assume $L = 0$. We need to show that $f(x) \to 0$ as $x \to \infty$. If $f(x)$ does not tend to 0 as $x \to \infty$, then there exists a sequence $\{x_n\}$ and $\varepsilon > 0$ where $x_n \to \infty$ as $n \to \infty$ such that $f(x_n) \geq \varepsilon$ (or $f(x_n) \leq \varepsilon$), $n = 1, 2, \cdots$. Suppose $f(x_n) \geq \varepsilon$. Since $f(x)$ is a slowly decreasing function, there exist $x_0 > 0$ and $\delta > 0$, such that

$$f(y) - f(x) > -\frac{\varepsilon}{2}$$

for $x \geq x_0$ and $0 \leq y - x \leq 2\delta$. Especially, let $x = x_n \geq x_0$. Since

$f(x_0) \geq \varepsilon$, $f(y) > \varepsilon/2$ for $x = x_n \geq x_0$ and $0 \leq y - x \leq 2\delta$. Thus we have

$$\begin{aligned}
&\frac{1}{\sqrt{2\pi}} \int_{-\infty}^{\infty} k_\lambda(x + \delta - t) f(t)\, dt \\
&\quad = \frac{1}{\sqrt{2\pi}} \left(\int_{-\infty}^{x} + \int_{x}^{x+2\delta} + \int_{x+2\delta}^{\infty} \right) k_\lambda(x + \delta - t) f(t)\, dt \\
&\quad \geq \frac{\varepsilon}{2\sqrt{2\pi}} \int_{x}^{x+2\delta} k_\lambda(x + \delta - t)\, dt - \frac{M}{\sqrt{2\pi}} \int_{-\infty}^{x} k_\lambda(x + \delta - t)\, dt \\
&\qquad\quad - \frac{M}{\sqrt{2\pi}} \int_{x+2\delta}^{\infty} k_\lambda(x + \delta - t)\, dt \\
&\quad = \frac{\varepsilon}{\sqrt{2\pi}} \int_{0}^{\delta} k_\lambda(x)\, dx - \frac{2M}{\sqrt{2\pi}} \int_{\delta}^{\infty} k_\lambda(x)\, dx \\
&\quad = \frac{\varepsilon}{\sqrt{2\pi}} \int_{0}^{\lambda\delta} \left(\frac{\sin x}{x} \right)^2 dx - \frac{2M}{\sqrt{2\pi}} \int_{\lambda\delta}^{\infty} \left(\frac{\sin x}{x} \right)^2 dx \to \frac{\varepsilon}{2}
\end{aligned}$$

as $\lambda \to \infty$. Therefore, there is a $\lambda_0 > 0$ (independent of x_n) such that

$$\frac{1}{\sqrt{2\pi}} \int_{-\infty}^{\infty} k_{\lambda_0}(x_n - t) f(t)\, dt > \frac{\varepsilon}{4}$$

for all $x_n \geq x_0$. This contradicts with the assumption and Lemma 8.3 follows.

Theorem 8.4 *(Ikehara Theorem)* *Suppose that $h(t)$ is a non-negative increasing function on $0 \leq t < \infty$ and the integral*

$$f(s) = \int_{0}^{\infty} \frac{h(t)}{e^{st}}\, dt \tag{8.35}$$

converges when $\operatorname{Re} s > 1$. Moreover, suppose that there exists a constant A such that

$$g(t) = \lim_{\sigma \to 1} \left(f(s) - \frac{A}{s - 1} \right), \quad s = \sigma + it \tag{8.36}$$

converges uniformly on every compact subset of $-\infty < t < \infty$ and $g'(t)$ is continuous on $-\infty < t < \infty$. Then

$$\lim_{t \to \infty} \frac{h(t)}{e^t} = A. \tag{8.37}$$

Proof Let

$$a(t) = \begin{cases} \dfrac{h(t)}{e^t}, & t \geq 0; \\ 0, & t < 0. \end{cases}$$

$$A(t) = \begin{cases} A, & t \geq 0; \\ 0, & t < 0. \end{cases}$$

If we can prove the following three facts:

(1) the integral

$$I_\lambda(x) = \frac{1}{\sqrt{2\pi}} \int_{-\infty}^{\infty} k_\lambda(x-t)(a(t) - A)\, dt$$

exists for any $\lambda > 0$;

(2)

$$\lim_{x \to \infty} I_\lambda(x) = 0;$$

(3) $a(t) - A(t)$ is a bounded slowly decreasing function,
then from Lemma 8.3, $a(x) \to A$ as $x \to \infty$, i.e. (8.36) holds.

Since the integral (8.34) converges on $\operatorname{Re} s > 1$, the integral

$$\begin{aligned} I_{\lambda,\varepsilon}(x) &= \frac{1}{\sqrt{2\pi}} \int_{-\infty}^{\infty} k_\lambda(x-t) \frac{a(t) - A(t)}{e^{\varepsilon t}}\, dt \\ &= \frac{1}{\sqrt{2\pi}} \int_{0}^{\infty} k_\lambda(x-t) \frac{a(t) - A(t)}{e^{\varepsilon t}}\, dt \end{aligned} \tag{8.38}$$

converges for $\lambda > 0$ and $\varepsilon > 0$. For any $\lambda > 0$,

$$\int_{-\infty}^{\infty} \frac{a(t) - A(t)}{e^{(\varepsilon + iy)t}}\, dt$$

converges uniformly on $|y| \leq 2\lambda$. Apply (8.28) to (8.37) we get that

$$\begin{aligned} I_{\lambda,\varepsilon}(x) &= \frac{1}{2\pi} \int_{-\infty}^{\infty} \frac{a(t) - A(t)}{e^{\varepsilon t}}\, dt \int_{-2\lambda}^{2\lambda} K_\lambda(y) e^{i(x-t)y}\, dy \\ &= \frac{1}{2\pi} \int_{-2\lambda}^{2\lambda} K_\lambda(y) e^{ixy}\, dy \int_{-\infty}^{\infty} \frac{a(t) - A(t)}{e^{(\varepsilon+iy)t}}\, dt \\ &= \frac{1}{2\pi} \int_{-2\lambda}^{2\lambda} K_\lambda(y) e^{ixy} \left(f(1 + \varepsilon + iy) - \frac{A}{\varepsilon + iy} \right) dy \end{aligned} \tag{8.39}$$

Since (8.35) converges uniformly on every compact subset,

$$\lim_{\varepsilon\to 0} I_{\lambda,\varepsilon}(x) = \frac{1}{2\pi}\int_{-2\lambda}^{2\lambda} g(y)K_\lambda(y)e^{ixy}\,dy \tag{8.40}$$

when we let $\varepsilon \to 0$ in the expression of $I_{\lambda,\varepsilon}(x)$. Also, since $g(y)k_\lambda(y)$ is continuous on $[-2\lambda, 2\lambda]$, apply (8.30) to (8.38) we get that

$$\lim_{x\to\infty}\lim_{\varepsilon\to 0} I_{\lambda,\varepsilon}(x) = 0. \tag{8.41}$$

On the other hand,

$$\begin{aligned}\lim_{\varepsilon\to 0} I_{\lambda,\varepsilon}(x) &= \lim_{\varepsilon\to 0}\frac{1}{\sqrt{2\pi}}\left(\int_0^\infty k_\lambda(x-t)a(t)e^{-\varepsilon t}\,dt - A\int_0^\infty k_\lambda(x-t)e^{-\varepsilon t}\,dt\right)\\ &= \frac{1}{\sqrt{2\pi}}\int_{-\infty}^\infty k_\lambda(x-t)(a(t)-A(t))\,dt = I_\lambda(x). \end{aligned} \tag{8.42}$$

Here we used the fact that if $f(t) \ge 0,\quad 0 \le t < \infty$, for any $T > 0$, $f(t)$ is piecewise continuous in $0 \le t \le T$, and for any $\varepsilon > 0$, the integral

$$\int_0^\infty e^{-\varepsilon t} f(t)\,dt$$

is convergent, then

$$\int_0^\infty e^{-\varepsilon t} f(t)\,dt = \int_0^\infty f(t)\,dt.$$

We leave the proof of this result to the reader. The existence of $I_\lambda(x)$ follows from (8.38) and (8.40). This proves (1). The fact (2) then follows from the equation (8.39). Now we need to show the fact (3).

By the definition of $A(t)$, we only need to show that $a(t)$ is a bounded slowly decreasing function. Since

$$\lim_{x\to\infty} I_\lambda(x) = 0,$$

we have

$$\begin{aligned}\lim_{x\to\infty}\frac{1}{\sqrt{2\pi}}\int_{-\infty}^\infty k_\lambda(x-t)a(t)\,dt &= \lim_{x\to\infty}\frac{1}{\sqrt{2\pi}}\int_{-\infty}^\infty k_\lambda(x-t)A(t)\,dt\\ &= \frac{A}{\sqrt{2\pi}}\lim_{x\to\infty}\int_{-\infty}^{\lambda x}\sqrt{\frac{2}{\pi}}\left(\frac{\sin u}{u}\right)^2 du\\ &= \frac{A}{\pi}\int_{-\infty}^\infty\left(\frac{\sin u}{u}\right)^2 du = A. \end{aligned} \tag{8.43}$$

Hence there exists x_0 such that

$$\frac{1}{\sqrt{2\pi}}\int_{-\infty}^{\infty} k_\lambda(x-t)a(t)\,dt < A+1$$

for $x > x_0$. That is

$$\int_{-\infty}^{\infty}\left(\frac{\sin t}{t}\right)^2 a\left(x-\frac{t}{\lambda}\right)dt < \pi(A+1)$$

for $x > x_0$. Since the integrand is non-negative, substitute $x + 2/\sqrt{\lambda}$ for x, we get that

$$\int_{-\sqrt{\lambda}}^{\sqrt{\lambda}}\left(\frac{\sin t}{t}\right)^2 a\left(x+\frac{2}{\sqrt{\lambda}}-\frac{t}{\lambda}\right)dt < \pi(A+1), \quad (x \geq x_0)$$

By the assumption, $e^t a(t)$ is an increasing function of t. So

$$a(x)e^{-\frac{3}{\sqrt{\lambda}}}\int_{-\sqrt{\lambda}}^{\sqrt{\lambda}}\left(\frac{\sin t}{t}\right)^2 dt < \pi(A+1)$$

for $x \geq x_0$. Let $\lambda \to \infty$. Then we have $a(t) \leq A+1$ for $x \geq x_0$. Since $h(x)$ is bounded for $x < x_0$, $a(x)$ is also bounded. Therefore, $a(x)$ is bounded on $-\infty < x < \infty$.

For any $\delta > 0$,

$$a(x+\delta) - a(x) = e^{-x}[e^{-\delta}h(x+\delta) - h(x)] \geq e^{-x}h(x)(e^{-\delta}-1).$$

Thus

$$\varliminf_{\substack{x\to\infty \\ \delta\to 0}} (a(x+\delta) - a(x)) \geq 0.$$

It follows that $a(x)$ is a slowly decreasing function and the theorem is proved.

Lemma 8.1 and Theorem 8.4 lead to the proof of Theorem 8.3

$$\lim_{x\to\infty}\frac{\psi(x)}{x} = \lim_{t\to\infty} e^{-t}\psi(e^t) = 1.$$

And the Prime Number Theorem follows.

Bibliography

Hua, L. K. [1] *Introduction to Higher Mathematics*, Vol. II (Science Press, Beijing, 1981) (in Chinese).

Hua, L. K. [2] *Introduction to Number Theory*, (Springer-Verlag, Berlin, Heidelberg, New York, 1982)

Wu, H. H., Lu, I. N. and Chen, Z. H. [1] *Introduction to Compact Riemann Surface* (Science Press, Beijing, 1981) (in Chinese).

Zhung, G. T. and Chang, N. O. [1] *Functions of Complex Variable* (Peking University Press, Beijing, 1984) (in Chinese).

Yu, J. Y. [1] *Functions of Complex Variable* (Higher Education Publishers, Beijing, 1980) (in Chinese).

Fan, L. L. and Wu, C. G. [1] *Theory of Functions of Complex Variable* (Shanghai Science and Technology Press, Shanghai, 1987) (in Chinese).

Gong, S. and Zhang, S. L. [1] *Concise Calculus* (University of Science and Technology of China Press, Hefei, 1997, 3rd Edition) (in Chinese)

Ahlfors, L. V. [1] *Complex Analysis*, 3rd ed. (McGraw-Hill, New York, 1979).

Ahlfors, L. V. [2] (1938). An extension of Schwarz's lemma, *Tran. Amer. Math. Soc. Proc.* **43**, pp. 359–364.

Ahlfors, L. V. [3] *Conformal Invariants* (McGraw-Hill, New York, 1973).

Ahlfors, L. V. and Sario, L. [1] *Riemann Surface* (Princeton University Press, Princeton, New Jersey, 1960).

Bell, S. R. [1] *The Cauchy Transform, Potential Theory and Conformal Mapping* (CRC Press, Boca Raton, Ann Arbon, London, Tokyo, 1992).

Berenstein, C. A. and Gay, R. [1] *Complex Variables, An Introduction* (Springer-Verlag, New York, Berlin, Heidelberg, London, 1991).

Cartan, H. [1] *Théorie Elementaire des Fonctions Analytique d'une on Plusieurs variables complexes* (Hermann, Paris, 1961).

Carathéodory, C. [1] *Conformal Representation* (Cambridge University Press, London, 1958).

Carathéodory, C. [2] *Theory of Functions of a Complex Variable*, Vol. I, II, (Chelsea, New York, 1954).

Cassou-Noguès, Ph. [1] *Elliptic Functions and Rings of Integers* (Birkhäuser, Boston, Basel, Stuttgart 1987).

Chandrasekharan, K. [1] *Elliptic Functions* (Springer-Verlag, Berlin, Heidelberg, New York, 1985).

Conway, J. B. [1] *Function of One Complex Variable* (Springer-Verlag, New York, Berlin, Heidelberg, London, 1986).

Fuchs, W. H. J. [1] *The Theory of Functions of One Complex Variable* (Van Nostrand, New York, 1967).

Grauert, H. and Reckziegel, H. [1] (1956). Hermiteschen metriken and normale familien holomorpher abbildungen, *Math. Z.* **89**: 108–125.

Greene, R. E. and Krantz, S. G. [1] (1987). *Biholomorphic Self-maps of Domains*, Lecture Notes, No. 1276, (Springer-Verlag, New York, Berlin, Heidelberg, London, 1987), pp. 136–207.

Greene, R. E. and Krantz, S. G. [2] *Function Theory of One Complex Variable*, A Wiley-Interscience Publication, John Wiley and Sons, Inc. New York, Chichester, 1997).

Heins, M. [1] *Complex Function Theory* (Academic Press, New York, 1968).

Hill, E. [1] *Analytic Function Theory* (Ginn and Company, Boston, 1959).

Hurwitz, A. and Conrant, R. [1] *Funktionen Theoie*, 4th ed. (Springer-Verlag, New York, Berlin, Heidelberg, London, 1964).

Jordan, C. [1] *Cour d'Analyse* (Gauthier-Villars, Paris, 1893).

Kobayashi, Z. [1] *Hyperbolic Manifolds and Holomorphic Mappings* (Marcel Dekker, New York, 1970).

Krantz, S. G. [1] (1990). *Complex Analysis; The Geometric Viewpoint* (MAA, Washington D. C. 1990).

Krantz, S. G. [2] *Function Theory of Several Complex Variables*, 2nd ed. (John Wiley and Sons, New York, 1991).

Krantz, S. G. [3] *Partial Differential Equations and Conformal Analysis* (CRC Press, Boca Raton, Ann Arbor, London, Tokyo,1992).

Lang, S. [1] *Complex Analysis*, 2nd ed. (Springer-Verlag, New York, Berlin, Heidelberg, London, 1985).

Lang, S. [2] *Elliptic Functions*, 2nd ed. (Springer-Verlag, New York, Berlin, Heidelberg, London, 1987).

Markushevich, A. I. [1] *Theory of Functions of a Complex Variable* (Chelsea, New York, 1977).

Minda, D. and Schober, G. [1] (1983). Another elementary approach to the theorems of Laudau, Montel, Picard and Schottky, *Complex Variables*, **2**, pp. 157–164

Narasimhan, R. [1] *Complex Analysis in One Variable* (Birkhäuser, Basel and Boston, 1984).

Narasimhan, R. [2] *Several Complex Variables* (University of Chicago Press, Chicago, 1971).

Nevanlinna, R. [1] *Analytic Functions* (Springer-Verlag, New York, Berlin, Heidelberg, London, 1970).

Palka, B. P. [1] *An Introduction to Complex Function Theory* (Springer-Verlag, New York, Berlin, Heidelberg, London, 1991).

Robinson, R. M. [1] (1939). A generalization of Picard's and related theorems, *Duke Math. J.* **5** pp. 118–132.

Rudin, W. [1] *Real and Complex Analysis* (McGraw-Hill, New York, 1966).

Saks, S. and Zygmund, A. [1] *Analytic Functions* (Elsevier, Amsterdam, London, New York, 1971).

Siegel, C. L. [1] *Topics in Complex Function Theory* (John Wiley and Sons, New York, 1969).

Springer, G. [1] *Introduction to Riemann surface* (Addison-Wesley Publishing Co., Massachusetts, 1957).

Titchmarsh, E. C. [1] *The Theory of Functions* (Oxford University Press, London, 1939).

Zalcman, L. [1] (1975). Aheuristic principle in complex function theory, *Amer. Math. Soc. Monthly* **82**, pp. 813–817.

Index